Cecil Hobart Peabody

Valve-gears for steam-engines

Cecil Hobart Peabody
Valve-gears for steam-engines
ISBN/EAN: 9783743467033
Manufactured in Europe, USA, Canada, Australia, Japa
Cover: Foto ©berggeist007 / pixelio.de

Manufactured and distributed by brebook publishing software (www.brebook.com)

Cecil Hobart Peabody

Valve-gears for steam-engines

VALVE-GEARS

FOR

STEAM-ENGINES.

BY

CECIL H. PEABODY,

ASSOCIATE PROFESSOR OF STEAM ENGINEERING AT THE MASSACHUSETTS
INSTITUTE OF TECHNOLOGY.

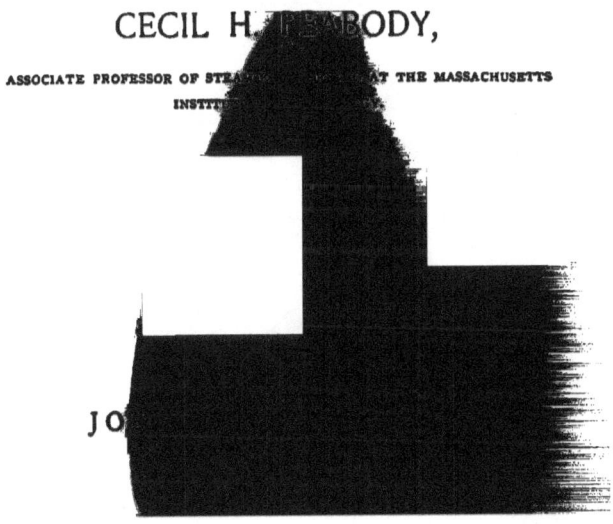

PREFACE.

THIS book is intended to give engineering students instruc-
[tio]n in the theory and practice of designing valve-gears for
[ste]am-engines. With the vast number of valves and gears in
[us]e at the present time, an exhaustive treatment in a text-
[bo]ok appears out of place; the author's aim is rather to give
[th]e learner a firm grasp of the principles and some facility in
[th]eir application. Each type discussed is illustrated by one or
[mo]re examples selected from good practice.

In the presentation of the elementary principles, geometri-
[ca]l or analytical methods are used as necessity or convenience
[ma]y suggest; but in the application, geometrical methods are
[us]ed exclusively, in conformity with the usual and preferable
[ha]bit of laying out valve-gears by construction. Zeuner's
[va]lve-diagram is used because it is widely and favorably known
[an]d appears to the author to be at least as good as any other
[cir]cular diagram.

In the discussion of radial valve-gears, the underlying
[pr]inciples found in all such gears are pointed out, and a few
[pr]ominent forms are illustrated. All such gears have neces-
[sa]rily or designedly large irregularities in their motions, so that
[an]alytical methods are useless if not misleading, and general·
[me]thods of treatment are of small value. Facility in design is
[to] be obtained through experience only.

Drop cut-off gears are represented by a few examples chosen to illustrate the principles and show a variety of treatment; especial attention is given to the Corliss gear. Advantage is taken in this connection of the opportunity to illustrate the use of cam-gears.

Common and well-known methods and processes have been used in most cases, and novelty has been rather avoided than sought. Some things, however, are believed to be here presented for the first time; for example, the combination of a skeleton model with construction for laying out link-motions and other irregular or complicated gears, and several examples of double valve-gears; the latter are introduced mainly to show the scope of the methods used. So much of the material used is the accumulation of common practice, and so many of the forms and methods are known by the names of the originators, that references to authorities or formal acknowledgments appear superfluous.

C. H. P.

MASSACHUSETTS INSTITUTE OF TECHNOLOGY,
March, 1892.

TABLE OF CONTENTS.

CHAPTER I.
PLAIN SLIDE-VALVE, ... 1

CHAPTER II.
SHIFTING ECCENTRICS, ... 33

CHAPTER III.
LINK-MOTIONS, ... 39

CHAPTER IV.
RADIAL VALVE-GEARS, ... 88

CHAPTER V.
DOUBLE VALVE-GEARS, ... 97

CHAPTER VI.
DROP CUT-OFF VALVE-GEARS, ... 115

VALVE-GEARS FOR STEAM-ENGINES.

CHAPTER I.

PLAIN SLIDE-VALVE.

THE valve-gear of a steam-engine consists of the valve, or valves, for admitting steam to, and exhausting steam from, the cylinder of the engine, together with the mechanism for giving motion to the valve, or valves. The discussion of valve-gears is therefore a part of kinematics or mechanism; the extent and importance of the subject make a separate presentation of it desirable.

The larger part of valve-gears derive their motion from one or more eccentrics; of such gears, the plain slide-valve is the simplest. Other valve-gears are best studied after an examination of the plain slide-valve, since they accomplish the same results, and by analogous methods.

Slide-valve Engine.—The working parts of a plain slide-valve engine are shown by Fig. 1, Pl. I; the frame is omitted in order that those parts may be the more readily understood. The centre of the main shaft is at O; the crank-pin is at C; the centre of the eccentric is at E; L is the connecting-rod, joining the crank-pin to the cross-head at H; l is the eccentric-rod, joining the eccentric to the head h of the valve-spindle. Both cross-head and valve-spindle head are constrained by guides to move in a line parallel to the axis of the cylinder. The steam-chest is represented with the cover off, to show the valve and its seat; the upper part of the valve is cut away to show the steam-ports and the exhaust-space. Fig. 2 gives a section of the valve and a half-section of the cylinder, with the piston P near the beginning of the stroke, and with the valve V partially open. The section being through the centre of the valve-spindle, the form of the exhaust-cavity is partially obscured. Fig. 3 gives a clearer idea of the valve and its seat.

Crank and Connecting-rod.—The crank, connecting-rod, and cross-head, in its guides, form what is called a slider-crank train, more clearly represented by OCH, Fig. 4.

Let the length of the connecting-rod be represented by L, the length of the crank by R, and the angle which the crank makes with the centre line XX', by θ; then the displacement HA of the cross-head from the beginning of its stroke is:

$$D = OA - Oc - (\overline{HC}^2 - \overline{Cc}^2)^{\frac{1}{2}};$$

$$D = L + R - R\cos\theta - (L^2 - R^2\sin^2\theta)^{\frac{1}{2}};$$

$$D = R(1 - \cos\theta) + L\left\{1 - \left(1 - \frac{R^2\sin^2\theta}{L^2}\right)^{\frac{1}{2}}\right\}. \quad (1)$$

In designing valve-gears, it is more convenient, and sufficiently accurate, to find the displacement of the piston for a given crank-angle by the construction shown by Fig. 4. The converse of this, i.e. to find the crank-angle corresponding to a given piston-position, is found with equal facility by such a construction, while the calculation by aid of equation (1) is troublesome. The principal use of that equation is in studying the nature of the irregularity introduced by the connecting-rod into the motion of the cross-head, and for that purpose it is convenient to expand the expression containing L by the

PLAIN SLIDE-VALVE. 3

binomial theorem, rejecting terms having the higher powers of L in the denominator; whence

$$D = R(1 - \cos \theta) + L\left\{1 - \left(1 - \frac{R^2 \sin^2 \theta}{2L^2}\right)\right\}$$

$$D = R(1 - \cos \theta) + \frac{R^2 \sin^2 \theta}{2L} \quad \ldots \ldots \quad (2)$$

The ratio of crank to connecting-rod in stationary-engine practice varies from 1 : 5 to 1 : 7½. In marine engineering the ratio 1 : 4, or even more, sometimes obtains. The maximum value of the term containing L occurs at $\theta = 90°$; for the ratios of crank to connecting-rod just given, the term containing L has then the values $\frac{1}{10}R$, $\frac{1}{15}R$, $\frac{1}{8}R$, respectively. It is apparent that the difference between the motion of the cross-head and piston, and harmonic motion represented by the equation

$$D = R(1 - \cos \theta), \quad \ldots \ldots \quad (3)$$

is always notable, and may be large.

Eccentric and Eccentric-rod.—The eccentric is derived from the crank, by the expansion of the crank-pin till it includes the shaft and obliterates the crank. Consequently the eccentric and eccentric-rod form a slider-crank train. The displacement of the valve is always reckoned from the middle position, and in Fig. 4 is $e = ha$; it may be calculated as follows:

$$e = Oe + Oa - (\overline{Eh}^2 - \overline{Ee}^2)^{\frac{1}{2}};$$

$$\therefore e = r \cos\{90° - (\theta + \delta)\} + l - [l^2 - r^2 \sin^2\{90° - (\theta + \delta)\}]^{\frac{1}{2}};$$

$$\therefore e = r \sin(\theta + \delta) + l\left[1 - \left\{1 - \frac{r^2 \cos^2(\theta + \delta)}{l^2}\right\}^{\frac{1}{2}}\right] \ldots \quad (4)$$

This equation differs from equation (1) only in that the eccentric-angle and the valve-displacement are reckoned from different points. Expanding by the binomial theorem and rejecting terms containing the higher powers of l in the denominator gives

$$e = r\sin(\theta + \delta) + \frac{r^2 \cos^2(\theta + \delta)}{2l}. \quad \ldots \quad (5)$$

The length of the eccentric-rod is commonly from 12 to 20 times the eccentricity; the right-hand term for such ratios will have the values $\frac{1}{24}r$ and $\frac{1}{40}r$, respectively, for maxima. It is customary to assume the motion of the valve to be harmonic, in which case it is represented by the equation

$$e = r\sin(\theta + \delta). \quad \ldots \ldots \quad (6)$$

The error of this assumption, though appreciable, is not large; moreover the method of setting valves of engines prevents any inconvenience from this source unless the eccentric-rods are very short.

The Slide-valve.—Fig. 3 gives the section of a plain slide-valve and its seat. The *ports* a, a_1 lead to the two ends of the cylinder; the *exhaust-space* s is connected with the exhaust-pipe; the *bridges* b, b_1 separate the ports from the exhaust-space. The steam-pressure in the steam-chest holds the valve against the seat and prevents leakage. The valve-seat is cut away so that the valve may over-travel its seat at the ends and thus both valve and seat may wear evenly. The edges of the ports and of the valves are machined and finished true; for convenience in the work, the edges of the ports and the inside edges of the valve are undercut as shown.

The valve commonly overlaps both inside and outside edges of the ports. The amount, o, by which the valve laps over the outside edge of the port is called the *outside lap* of the valve;

in like manner, *i* is called the *inside lap*. Frequently the outside lap is called simply the *lap*. In the figure the laps are equal; in some cases they are unequal. The inside lap may be nothing, or negative, in which case the valve is said to have a *clearance*.

In Fig. 2 the valve is shown admitting steam to the end of the cylinder remote from the crank, called the head end; and it is exhausting steam from the other end of the cylinder, called the crank end. The valve is set in such a manner that when the engine is on a dead-point the valve is open by a small amount called the *lead*. For this purpose the eccentric is set ahead of the crank 90° plus the angle δ, called the *angular advance*. As the crank moves forward the valve opens more and more till the centre of the eccentric comes to the line of centres, then the valve begins to return and it shuts the port before the stroke of the piston is finished. The stroke from the head to the crank end is called the *forward stroke;* the *return stroke* is made from the crank to the head end. The valve is open by the amount of the lead at the beginning of the return stroke, and at the same time the exhaust is open for the head end. As in the forward stroke, the valve first opens wider as the stroke proceeds, then it returns and closes the port before the piston reaches the end of the stroke.

Events of the Stroke.—When the outside edge of the valve is at the edge of the port, as shown by Fig. 4, Pl. II, and the valve is opening, *admission* is said to occur. This happens just before the stroke commences. When the valve is in the same position, but is closing, *cut-off* takes place. At either position the displacement of the valve is equal to the outside lap; consequently we have the following principle:

When the displacement of the valve is equal to the outside lap, the engine is either at admission or cut-off.

Release occurs when the inside edge of the valve is at the edge of the port and the valve is opening to exhaust, as shown by Fig. 5, Pl. II.

Compression occurs when the valve is at the same position but is closing the exhaust. At either position the displacement of the valve is equal to the inside lap; consequently we have:

When the displacement of the valve is equal to the inside lap, the engine is either at release or at compression.

Valve-ellipse.—The action of the valve is conveniently studied by the aid of a diagram like that shown by Fig. 1, Pl. II, and known as a valve-ellipse. It is obtained by laying off the displacements of the piston (HA, Fig. 4, Pl. I) as abscissæ, and the displacements of the valve (ha, Fig. 4, Pl. I) as ordinates, for a sufficient number of crank-positions, and by drawing a smooth curve through the points thus found. Any convenient scale may be used for either abscissæ or ordinates; in practice the ordinates may be full scale, while the abscissæ are quarter scale or less. In the figure the abscissæ are made equal to the displacements of the piston from Fig. 4, Pl. I, and the ordinates are made twice the valve-displacements from the same figure.

The figure is properly an oval on account of the influence of both connecting-rod and eccentric-rod. The dotted ellipse is drawn to show the irregularities, which are large in the figure on account of the unfavorable ratios of crank to connecting-rod, and of eccentric to eccentric-rod, in Fig. 4, Pl. I.

The lines oo', o_1o_1' are drawn parallel to AA', and at a distance $Ao = Ao_1$ equal to the outside lap; and the lines ii'', i_1i_1' are at a distance $Ai = Ai_1$ equal to the inside lap. Now a displacement of the valve equal to the outside lap will give either admission or cut-off; an inspection of the diagram shows that admission occurs at 0.02 of the stroke before the dead-point, and that cut-off occurs at 0.88 of the stroke, for the forward stroke; for the return stroke, the admission occurs just before the dead-point, and cut-off occurs at 0.72 of the stroke. Compression occurs at 0.94 and release at 0.98 of the stroke, for the forward stroke; and for the return stroke compression occurs at 0.83 and release at 0.93 of the stroke. The lead for the forward stroke is on, and for the return stroke $o_1'n'$. The inequality of the leads is due to the short eccentric-rod; in practice such an inequality does not occur, first, because the eccentric-rod is longer, and second, because the usual method of setting the valve makes the lead equal.

The valve-ellipse may be advantageously used to investigate the action of a valve having an irregular motion, such as is given by some special valve-gears to be studied later, and it should be drawn during the design of the valves for every important engine. The motion of valves of an existing engine may be investigated by causing the engine to draw its own valve-ellipse. For this purpose, a reduced copy of the piston-motion, obtained by aid of a pantograph or otherwise, may be communicated to a slip of paper on which is pressed a pencil that derives its motion across the slip of paper from the valve-gear. The oval thus drawn will have the piston-displacements for abscissæ and the valve-displacements for ordinates, and should be identical with the valve-ellipse drawn to the same scale; any discrepancy must be due to mechanical defects in the valve-gear.

Sinusoidal Diagram.—A diagram called by this name, because the curves resemble sinusoids, was devised by Moll and Montéty for use in designing valve-gears, taking account of the irregularities of both the piston and the valve. Starting at A, Fig. 2, the crank-angles are laid off as abscissæ toward A''; and both the piston-displacement and the valve-displacement for a given crank-angle are laid off as ordinates, thus giving two curves, $AA'''A''$, representing the piston-motion, and $nn'n$, representing the valve-motion. The dotted lines are true sinusoids, and would represent the piston and valve motions if both were harmonic. The lines $oo'o$, $o_1o_1'o_1$ are drawn to represent the outside laps, and the lines $ii'i$, $i_1i_1'i$, to represent the inside laps, which may or may not be equal. Inspection of the diagram shows that cut-off occurs at the crank-angle 133°, and at a piston displacement equal to *ab*. Conversely, if the cut-off is desired to take place at the piston-position *ab*, draw the line $b'b''$ parallel to AA'' and at a distance from it equal to the desired piston-displacement; from *b*, the intersection with the curve of piston-displacements, draw the ordinate *ab*; then *ac* is the lap which will give the desired cut-off. It is convenient to draw the curve of piston-displacements on a sheet of paper on a drawing-board, and to draw the curve of valve-displacements, which may be extended to give about one and a half revolutions, on a piece of tracing-paper or tracing-cloth. By superimposing the tracing of the valve-displacements on the drawing of the piston-displacements, and slipping it along as desired, the effect of using different values for the angular advance may be readily determined; at the same time the effect of different laps may be determined, or the lap for a special purpose may be found.

This diagram cannot be conveniently substituted for the valve-ellipse, since it does not present to the eye the character of the valve-motion combined with the piston-motion; a valve-ellipse can readily be drawn from the sinusoidal diagram.

Zeuner's Diagram.—Fig. 7 shows a diagram devised by Zeuner for investigating the action of, and for designing, valves which have harmonic motion. Let XOX' and OY be a pair of rectangular axes, and let the crank have a left-handed rotation as shown by the arrow. Lay off the angle $YOP = \delta$ equal to the angular advance, *toward* the crank; make OP equal to the eccentricity, and on it draw a circle ONP called the valve-circle; then the valve-displacement for a given crank-angle θ is equal to the chord ON, cut off by the valve-circle from the line representing the crank. To prove this, we have Op for the position of the eccentric corresponding to the crank-position OR, obtained by making the angle pOR equal to $90° + \delta$; and we have On for the valve-displacement on the assumption of harmonic motion, and

$On = r\cos pOn = r\cos(180° - 90° - \theta - \delta) = e$;

∴ $e = r\sin(\theta + \delta)$,

as given by equation (6). But the triangles *Opn* and *OPN* are equal, since they are right-angled triangles with the sides *Op* and *OP* equal, and the angles *PON* and *pOn* are each equal to $180° - 90° - \theta - \delta$.

The use of the diagram is shown by Fig. 1, Pl. III. Two valve-circles are drawn, *OP* for the forward and *OP'* for the return stroke. The circles $oo'o''$ and $ii'i''$, drawn with the outside and inside laps as radii, are called the lap-circles. At the beginning of the stroke the valve-displacement is $On = Oo + on$, and the lead is *on*; the valve is in the position shown by Fig. 3, Pl. II. As the crank moves forward the valve opens, at first rapidly, and then more slowly, till the maximum displacement is attained, when the crank coincides with *OP*; the position of the valve is then shown by Fig. 6, Pl. II. As the crank moves beyond the last position the valve returns, till at OR_c the displacement is equal to the lap and cut-off occurs; the position of the valve is then shown by Fig. 4, Pl. II, and it is moving toward the right to shut the port. At OR_k the displacement of the valve is equal to the inside lap and compression occurs; the position of the valve is shown by Fig. 5, Pl. II, and it is moving toward the right to shut the exhaust-port. At OR_r the displacement of the valve is again equal to the inside lap and release occurs; the valve now has the left-hand inside edge on the edge of the port and it is moving to open the port. At OR_a' the displacement of the valve is equal to the outside lap and admission occurs at the crank end; the position of the valve to give admission at the head end (corresponding to the crank-position OR_a) is shown by Fig. 4, Pl. II. It is customary to associate the admission at OR_a, in anticipation of the forward stroke, with that stroke, and to associate the admission at OR_a' with the return stroke.

Though it is not customary to do so, the two arcs $tt't''$, $ss's''$ may be added to find the crank-positions at which the edge of the valve is on the further edge of the port, and the port is wide open. The radius Ot' is made equal to the width of the port plus the outside lap; when the crank-position passes through t or t'' the displacement of the valve is equal to the lap plus the width of the port and the port is then wide open. In like manner the radius Os' is made equal to the width of the port plus the inside lap, and when the crank-position passes through s or s'' the port is wide open for exhaust.

The diagram shows that the outside edge of the valve over-travels the edge of the port by the amount $t'P$. Some over-travel is desirable to give a free flow of steam to the cylinder and a rapid admission and a sharp cut-off. The over-travel of the valve for exhaust is $s'P$, which is greater than $t'P$ by the amount of the difference of the outside and inside laps. The amount by which the port is open at any position of the valve is called the port-opening. The maximum port-opening for supply, which occurs when the crank coincides with OP, is equal to $o'P$, the difference between the eccentricity and the lap. The maximum port-opening for exhaust is equal to $i'P$. A slide-valve moved by an eccentric always has the maximum port-opening for exhaust at least as great as the width of the port, and it is commonly greater; the maximum port-opening for supply is also commonly greater than the width of the port, but it is sometimes less. A slide-valve moved by a gear that gives a variable cut-off, as will be seen later, may have both maximum port-openings less than the width of the port for some settings of the gear.

An inspection of the diagram Fig. 1, Pl. III, will show that a change of the outside lap will affect both admission and cut-off; thus, the cut-off is hastened and the admission is delayed by an increase of the outside lap, and conversely the admission comes earlier and the cut-off comes later if the outside lap is decreased. In a similar way, increasing the inside lap delays the release and hastens the compression, while decreasing that lap produces a contrary effect. Since it is the relative proportions of lap and eccentricity which determine the cut-off, it is apparent that decreasing the eccentricity with a constant lap produces the same effect as increasing the lap with a constant eccentricity; i.e. it hastens the cut-off and delays the admission. Finally, it will be seen that increasing the angular advance hastens all the events of the stroke; and that decreasing the angular advance delays all the events of the stroke.

Should the inside lap be made nothing, so that in mid-gear the inside edge of the valve shall coincide with the edge of the port, then both compression and release will occur when the crank is at right angles to POP'. Sometimes, in designing or remodelling a valve, it will be found advisable to give a *clearance* to the valve instead of an inside lap, in which case the engine will for a short time, near the end of the stroke, exhaust from both ends at once. Suppose that the circle $ii'i''$, Fig. 1, Pl. III, represents a clearance, then release will occur at OR_k and compression at OR_r.

Expansion and Compression.—From cut-off at OR_c (Fig. 1, Pl. III) to release at OR_r, the head end of the cylinder is shut off from both the supply and the exhaust. While the crank moves forward from R_c to R_r the piston moves a corresponding amount, and the steam in the cylinder expands and experiences a loss of pressure in consequence; this action is called the expansion. When the valve closes the exhaust at OR_k the steam then caught in the cylinder is compressed ahead of the piston till a new supply of steam is admitted at OR_a'; this action is called the compression. On the return stroke, in a like manner, steam is expanded from OR_c' to OR_r', and is compressed from OR_k' to OR_a.

Rocker and Bell-crank Lever.—In the work thus far it has been assumed that the centre-lines of the crank and connecting-rod with the cross-head and piston, and of the eccentric and eccentric-rod with valve-spindle and valve, coincide in the elevation, as shown by Fig. 1, Pl. I. This assumption is convenient in giving the first description of the valve-gear and in discussing the action and the theory of the valve-motion; and the design of the valve is commonly carried on as though such a coincidence existed, the deviation from such a coincidence being considered only in the mechanical problem of laying out the mechanism of the engine.

If the centre-line of the valve-spindle passes through the centre of the shaft, and the eccentric-rod is connected directly to the valve-spindle, then the motion of both crank and connecting-rod, and eccentric and eccentric-rod, referred to their own proper axes, will be the same as already found, even though their centre-lines do not coincide. Such a lack of coincidence will make the angle between the eccentric and the crank more (or less) than 90° plus the angular advance, by the amount of the angle between the two centre-lines. This difference needs consideration only in the process of setting the valve; and if the angle between the centre-lines is small, it will require little or no attention at that time. Since such an arrangement involves a lack of parallelism between the paths of the valve and of the piston, the work of boring thecylinder and facing off the valve-seat is more troublesome, and other machine-work is more difficult, unless special processes are provided; consequently this arrangement is seldom adopted.

Very commonly the paths of the piston and the valve are parallel but do not coincide in the elevation; thus, the axis of the crank and connecting-rod with the cross-head and cylinder may be XX' in the Figs. 2, 3, and 4, Pl. III, while the centre-line of the valve-spindle may be xx' in the same figures. In such cases a rocker or a bell-crank lever should be used to transmit motion from the eccentric to the valve.

The following method may be used in laying out a bell-crank lever. Let A be the position of the end of the valve-spindle when the valve is in mid-position; lay off $Aa = Aa'$ equal to the lap of the valve, and with a radius equal to the length of the arm of the bell-crank lever draw arcs intersecting at C; this gives the axis about which the bell-crank lever vibrates. This construction prevents the bell-crank lever from introducing any irregularity into the action of the valve at admission and cut-off; irregularities at other times are of less consequence. In laying out such a gear for a locomotive with a rigid valve-spindle that extends directly from the end of the bell-crank lever or rocker to the valve-yoke, it is important to have the bending or lateral motion of the valve-spindle as small as possible; in such case the point C may be so chosen that the lateral motion of the end of the valve-spindle shall be half above and half below the line xx'. With a radius equal to BC, the other arm of the bell-crank lever, draw an arc as shown, and draw a tangent to this arc from O. Draw perpendiculars OE and CB from O and C to this tangent; then EB is the length of the eccentric-rod. If desired, the relation of the crank and eccentric may be found by laying off the angle $XOR = \delta$, the angular advance, since the crank is at that angle before the dead-point when the valve is in mid-position; the angle EOR will not be equal to $90° + \delta$, but this is a matter that affects the valve-setting only, and even in that process the exact knowledge of the angle between the crank and eccentric is not of importance.

If it is considered of importance that the eccentric-rod shall be some definite length, then the centre C, on which the bell-crank lever vibrates, may be shifted so as to give that length. If C is to be shifted a short distance, then a line parallel to XX' may be drawn through B, and with a radius equal to the desired length of the eccentric-rod an arc may be drawn from E intersecting that parallel line at a point B'; the whole bell-crank lever is to be shifted bodily to the extent BB', and the length of the valve-spindle must be changed the same amount.

In the figure the arm CB is made $\frac{3}{4}$ of CA; consequently the motion of the valve will be that which would be given by an eccentric equal to $\frac{3}{4}$ of OE if the connection were direct.

In designing and laying out the valve it is treated as though it were moved by such an eccentric. The ratio of the arms may be made anything desired; they have commonly the same length.

In laying out a rocker, the process is the same as that just described for the bell-crank lever, except that the arm CB, Fig. 3, is laid off on the side opposite A, and the eccentric follows the crank.

Fig. 4 shows the equal-armed straight rocker with the centre of vibration C midway between the lines XX' and xx'; it may be made a little longer so as to give a construction equivalent to that shown by aAa', Figs. 2 and 3.

Area of Steam-pipe and Steam-ports.—In order that the loss of pressure in the steam-pipe due to friction may not be excessive, it is customary to limit the velocity to 100 feet a second or 6000 feet per minute. The volume of steam supplied to an engine is calculated, for this purpose, on the assumption that the cylinder is filled at each stroke.

For example, the diameter of the steam-pipe of an engine having a diameter of 18 inches and a stroke of 3 feet, and making 75 revolutions per minute, should be 5 inches. It may be found as follows: The piston displacement is

$$\frac{\pi}{4}\left(\frac{18}{12}\right)^2 \times 3 = 1.7671 \times 3 = 5.30 \text{ cubic feet};$$

the volume of steam per second is

$$2 \times 5.30 \times \frac{75}{60} = 13.25 \text{ cubic feet};$$

the area of the steam-pipe should be

$$\frac{13.25}{100} \times 144 = 19.08 \text{ square inches},$$

and the diameter should be

$$\left(\frac{4}{\pi} \times 19.08\right)^{\frac{1}{2}} = 4.93 \text{ inches,}$$

or nearly 5 inches.

Even though the cut-off occurs before the end of the stroke and the actual volume of steam used is less than the volume calculated by the above method, the area of the steam-pipe should not be reduced, for the rate of the flow of steam will be as great when the piston is near mid-stroke. The steam-pipe supplying an engine which has an early cut-off (one third stroke or less) may be made less than given by the above method, provided there is a large steam-chest, or a steam-drum near the engine. Seaton[*] states that a velocity of 8000 feet may be allowed in the steam-pipe of a marine engine; and that 10,000 feet may be allowed for very large engines.

The area of ports and passages leading to the cylinder should be equal to that of the steam-pipe; and the area of ports and passages leading from the cylinder should be double that area. If the steam-pipe is calculated on the assumption of a velocity of 6000 feet a minute, this rule will often make the size of ports and passages such that it is difficult, if not impossible, to provide for them in the design of a high-speed engine. In such case the area of ports and passages must be reduced, but they should never be less than half the area given by that rule. This is equivalent to calculating the area of a steam-port by the method given for a steam-pipe, except that a velocity of 12,000 feet a minute is allowed; and then making the exhaust-passage twice that area. Since both supply and exhaust pass through the ports of a plain slide-valve engine, the area must be made sufficient for the exhaust.

For example: the engine mentioned above should have the area of the port

$$2 \times 19.08 = 38.16;$$

---

[*] Manual of Marine Engineering.

and if the length of the port is made eight tenths of the diameter, or 14 inches, the port will be

$$38.16 \div 14 = 2.65 \text{ inches.}$$

If the area of the port is reduced to half as much the width will become $1\frac{3}{8}$ inches.

When it becomes difficult or undesirable to give the slide-valve sufficient motion to open the steam-port wide for the supply, the maximum port-opening may be made from $\frac{6}{10}$ to $\frac{9}{10}$ the width of the port.

When special valve-gears are used that open the valve rapidly and close it promptly, the area of the ports and passages may be made smaller than the above methods provide, but such reduction should be made only with complete knowledge of the action of the valve and of the effect of the reduction of the flow of steam.

Lead and Lead-angle.—The lead, or the amount that the valve is open when the engine is on a dead-point, varies with the type and size of the engine, from a very small amount (or even nothing) up to $\frac{3}{8}$ of an inch or more. Stationary engines running at slow speed may have from $\frac{1}{64}$ to $\frac{1}{16}$ of an inch lead. The effect of compression is to fill the waste space at the end of the cylinder with steam; consequently engines having much compression need less lead. Locomotive-engines having the valves controlled by the ordinary form of Stephenson link-motion may have a small lead when running slowly and with a long cut-off, but when running at speed with a short cut-off the lead is at least $\frac{1}{4}$ of an inch; and locomotives that have a valve-gear which gives constant lead commonly have $\frac{1}{4}$ of an inch lead.

Zeuner's diagram does not admit of the use of the lead in solving problems that arise in designing valves, but we may use instead the *lead-angle*, or the angle that the crank makes with the line of dead-points at admission; in Fig. 1, Pl. III,

XOR_a is the lead-angle. The lead-angle may vary from zero to 8°; a convenient lead-angle for solving problems is $2\frac{1}{2}°$. After the required problem is solved by the aid of the lead-angle the conditions may be varied so as to give a desired lead.

Problems on the Slide-valve.—By assuming various elements of the valve to be known, a series of problems relating to the plain slide-valve may be stated and solved. Of such problems *one* has a real importance to the designer of valve-gears; others are either so simple as to require no formal solution, or they are problems that are not likely to arise in practice. This important problem is given below as Problem II; the other, Problem I, is given as a convenient introduction to it, for students who here approach the subject for the first time.

PROBLEM I. *Given the eccentricity, the lead-angle, and the crank-angle at cut-off, to find the angular advance, the lap, and the lead.*

In Fig. 1, Pl. IV, draw, to any convenient scale, the arc XR_cX' to represent the path of the crank, referred to the axes XOX' and OY. Lay off the angle XOR_a equal to the lead-angle, and lay off XOR_c, the crank-angle at the point of cut-off. Should the piston-position at cut-off be given instead of the crank-angle, lay off Or, the given piston-position at cut-off, and draw the vertical line rR_c, to find the crank-position OR_c corresponding. With equal leads and laps the crank-angle at cut-off will be the same for each stroke, and the mean piston-position will be nearly equal to the piston-position at cut-off with harmonic motion; hence the above construction.

Bisect the angle R_aOR_c and draw the line OP; on it draw the valve-circle $OoPo'''$, and through the intersections of this valve-circle by the lines OR_a and OR_c draw the lap-circle $oo'o''$. The lead is Oa.

In Fig. 1 the eccentricity is $1\frac{1}{2}$ of an inch, the cut-off is at $\frac{3}{4}$ of the stroke, and the lead-angle is $2\frac{1}{2}°$. The lap is $\frac{23}{32}$ of an inch, and the lead is $\frac{1}{16}$ of an inch; the angular advance is $31\frac{1}{4}°$.

PROBLEM II. *Given the crank-angle at cut-off, the lead-angle, and the maximum port-opening, to find the eccentricity, the lap, and the angular advance.*

As has already been stated, this problem is the one met by the designer in laying out a valve-gear. The angular advance is obtained by the same process as in Problem I; namely, by bisecting the angle R_aOR_c.

Now assume a trial eccentricity, preferably a little larger than the required eccentricity, and draw an assumed valve-circle OqP_1, and the corresponding lap-circle $qq'q''$. The maximum port-opening with the assumed valve-circle is $q''P_1$; and from it the diameter of the required valve-circle, equal to the required eccentricity, may be obtained by the proportion

Assumed port-opening : actual port-opening ::

Assumed eccentricity : required eccentricity.

A graphical solution may be made by aid of similar triangles, as follows: Lay off the line $Oc_1 = q''P_1$, in a convenient position, and join C_1c_1; make Oc equal to the given maximum port-opening, and draw cC parallel to c_1C_1; then C is the centre of the required valve-circle and OP is the required eccentricity. The lap-circle $oo'o'''$ is drawn through the intersections of the valve-circle by R_a and R_c.

In Fig. 2, Pl. IV, the angular advance is $31\frac{1}{4}°$, as in Fig. 1; the assumed eccentricity is $1\frac{3}{4}$ of an inch, which gives a port-opening of $\frac{28}{32}$; the required maximum port-opening is taken to be $\frac{3}{4}$ of an inch; the required eccentricity is $1\frac{7}{16}$ of an inch; and the lead is $\frac{1}{16}$ of an inch, or a trifle less.

Modifications.—In general it is not of great importance that the cut-off shall occur exactly at the chosen crank-angle or piston-position, and it is seldom necessary to know the angular advance except in drawing the valve-diagram. It is, however, very convenient if not necessary that the lead and lap shall be some determined quantity stated in fractions of an inch that are commonly used in the machine-shop. By judicious modifications the designer may secure this for the valve without seriously affecting the point of cut-off. In Fig. 3, Pl. IV, the lap, Oo, is made $\frac{3}{4}$ of an inch, and the lead is $\frac{1}{32}$ of an inch. At a the vertical aP is drawn, and from O, with a radius equal to $1\frac{1}{2}$ of an inch, the vertical is intersected in P; thus giving the diameter OP on which the valve-circle $OaPo''$ is drawn. The cut-off comes at the crank-position OR_c, corresponding to the piston-position Xr instead of $Xt = \frac{3}{4}$ stroke, as required.

The process of laying out a slide-valve will be considered in connection with the valve for equal cut-off.

Equalization of Cut-off at the Expense of the Lead.— Let it be assumed that the cut-off shall occur at a given piston position for each stroke, taking account of the irregularity due to the connecting-rod. Draw the line xX', Fig. 1, Pl. V, on which choose O for the centre of the crank and for the origin of coördinates, and draw the vertical axis YOY'. Draw the circle $XYX'Y'$, to any convenient scale, to represent the crank-circle. With a radius equal to the length of the connecting-rod, on the same scale, and with X and X' as centres, cut the centre-line $X'x$ at x and x'; this will give the stroke of the cross-head, equal to the stroke of the piston. Lay off the point on the stroke at which cut-off is to occur on both strokes, and with these points as centres and the length of the connecting-rod as a radius intersect the crank-circle at R_c and R_c'. In the figure the connecting-rod is taken to be five times the crank, and cut-off is assumed to occur at $\frac{3}{4}$ of the stroke. It is at once apparent that the crank-angles XOR_c and $X'OR_c'$ are not equal, and OR_c and OR_c' are not one straight line. Choose a small head-end lead-angle XOR_a; in the figure it is 1°. Bisect the angle R_aOR_c, and draw the line POP', on which draw the two valve-circles as shown. The eccentricity may be determined by a preliminary solution, assuming harmonic motion, as in Problem II, or the same solution may be made directly on the figure;

if the latter method is used, confusion is liable to occur from the number of circles drawn, especially if the eccentricity is modified to get some convenient dimension; consequently it is better to make such a construction separately and transfer the results to the main diagram. Draw the lap-circle oo for the upper valve-circle, through the intersections of that circle by OR_a and OR_c; and draw the lap-circle $o'o'$ for the lower valve-circle, through the intersection OR_c' with that circle. The admission at OR_a' occurs at the intersection of the lower valve-circle and the lap-circle $o'o'$; the crank-end lead is large, if not excessive.

It is customary, in designing a valve for equal cut-off, to equalize the compression also. In Fig. 1 the compression is assumed to occur at $\frac{7}{8}$ of the stroke, or at the crank-positions OR_k and OR_k'. The inside laps are ii and $i'i'$, so that the release occurs at R_r and R_r'. The point of intersection i of the line OR_k with the valve-circle may be determined by dropping a perpendicular Pi from P on OR_k. The valve-diagram in Fig. 1 gives the following dimensions:

$$\begin{aligned}
&\text{Eccentricity} \ldots\ldots\ldots\ldots\ldots\ldots\ldots\ldots 1\tfrac{1}{2} \text{ inch.}\\
&\text{Outside lap, forward stroke} \ldots\ldots\ldots\ldots \tfrac{51}{64} \text{ "}\\
&\qquad\qquad \text{return stroke} \ldots\ldots\ldots\ldots \tfrac{37}{64} \text{ "}\\
&\text{Inside lap, forward stroke} \ldots\ldots\ldots\ldots \tfrac{6}{16} \text{ "}\\
&\qquad\qquad \text{return stroke} \ldots\ldots\ldots\ldots \tfrac{9}{64} \text{ "}\\
&\text{Head-end lead} \ldots\ldots\ldots\ldots\ldots\ldots\ldots \tfrac{1}{64} \text{ "}\\
&\text{Crank-end lead} \ldots\ldots\ldots\ldots\ldots\ldots\ldots \tfrac{15}{64} \text{ "}
\end{aligned}$$

To lay out a Slide-valve.—The valve for which dimensions were found in Fig. 1 Pl. V, is shown in section by Fig. 2. To lay out the valve, begin at the crank end and make $ab = \tfrac{37}{64}$ of an inch, equal to the return-stroke outside lap; make $bc = \tfrac{9}{16}$ of an inch, equal to the width of the port.

The greatest displacement of the valve, equal to the eccentricity $1\tfrac{1}{2}$ of an inch, will carry the point a to a', and when the valve is in that position it must not overrun the edge of the bridge, but rather there must be width enough remaining to prevent leakage. The least width of bridge in the figure is $\frac{5}{8}$ of an inch, and the width of $\frac{1}{2}$ inch is chosen to insure a joint.

The forward-stroke inside lap, $\frac{5}{16}$ of an inch, is laid off at cd. The greatest displacement of the valve will carry the point d to d', and at that position of the valve the remnant of the exhaust-space should be at least as wide as the port, i.e. $\frac{9}{16}$ of an inch as shown. The exhaust-space is commonly made wider than this construction gives; it should not be unduly increased, since it will then make the valve large and the friction excessive.

The valve is completed by making the width of the bridge $\frac{1}{2}$ of an inch and of the port, $\frac{9}{16}$ of an inch, as shown. If the eccentricity, $1\frac{1}{2}$ of an inch, be laid off toward the left, from the right-hand edges of the valve, it will appear that the right-hand bridge is wider than necessary, and that the remnant of the exhaust-space, when the valve has its maximum displacement to the left, is greater than the width of the steam-port. No inconvenience will occur from such an excess of bridge or exhaust-space; but had the construction been begun at the right hand, then both the bridge and the exhaust-space would be too narrow. For constructive reasons, the bridge for any slide-valve may be made wider than required to prevent leakage.

Fig. 3, Pl. V, gives the section (half-size) of a valve with equal laps which will give the same average cut-off as the valve shown by Fig. 2; the cut-off at the head end will be longer, and that at the crank end will be shorter. The outside lap for Fig. 3 is very nearly the mean of the unequal outside laps of Fig. 2; and the inside lap is very nearly the mean of the inside laps of the same figure. Such a valve may be laid off, beginning at either end.

The height of the exhaust-cavity of the valve should never be less than the width of the steam-port; it is commonly once and a half as high, if not more.

The method just given for laying out the slide-valve has

the apparent inconvenience that the centre of the exhaust-space cannot be directly located on the assembly drawing of the engine. This difficulty is, however, only apparent, for the section of the valve is commonly drawn separately and at full size, and then can be transferred to the assembly drawing, which may be to any convenient scale. The results obtained by laying out the work on the drawing-board should always be checked by a numerical calculation; and if desired such a numerical calculation may be made first, but it should be checked by the subsequent laying out of the valve. Thus the width of the bridge should be greater than

$1\frac{1}{2} - \frac{37}{64} - \frac{9}{16} = \frac{23}{64}$, or nearly $\frac{3}{8}$ of an inch;

and the exhaust-space should have a width of

$\frac{5}{16} + 1\frac{1}{2} + \frac{9}{16} - \frac{1}{2} = 1\frac{7}{8}$ of an inch.

Equalization of Cut-off with Rocker.—A method of equalizing the cut-off without destroying the equality of the lead was devised by Professor Sweet, and is employed on the Straight Line engine. The same method may be employed with a bell-crank lever; the construction must, however, be carried out for each case separately.

On Plate VI the centre-line of the crank and connecting-rod is xX', while the centre-line of the valve-spindle is nn'. Assuming that the cut-off is to occur at $\frac{3}{4}$ of the stroke, divide the diameter XX' of the crank-pin circle into fourths and draw the vertical line NR_o from N; in the figure the eccentric-circle, drawn to a larger scale than the crank-circle, happens to pass through N. Make the angle XOR_a equal to the lead-angle, and bisect the angle R_aOR_c by the line OP. Draw the valve-circle on OP and the lap-circle oo', thus finding the outside lap.

With C and C', at $\frac{3}{4}$ of the forward and return strokes as centres, and with the length of the connecting-rod as radius, intersect the crank-circle at R_c and R_c', to find the crank-positions at cut-off. Produce the line R_aO to R_a', the crank-position at admission for the return stroke.

Since the gear is drawn with a rocker, the eccentric will follow the crank at an angle equal to XOP. The positions, r_c and r_c', of the eccentric at cut-off will be found by laying off the angles R_cOr_c and $R_c'Or_c'$, equal to XOP; and the positions of the eccentric, r_a and r_a', at admission are found by making the angles R_aOr_a and $R_a'Or_a'$ equal to the same angle.

With r_a and r_c as centres, and with a radius equal to the length of the eccentric-rod, draw two arcs intersecting at e_1; also with the same radius draw arcs from r_a' and r_c' intersecting at e; then if the end of the eccentric-rod be guided on a path passing through e and e_1 the lead and cut-off will both be equalized. With a radius equal to the length of the arm of the rocker, draw arcs intersecting at T_1, and from T_1 draw an arc through ee_1, and also the chord ee_1; the length of the chord is greater than twice the lap; consequently the other arm of the rocker should be made smaller than T_1e, in proportion as the lap is less than half ee_1. The length of the other arm may be found by a numerical calculation, or by the following construction: make qs equal to the lap, and draw st parallel to eT_1; then st is the required length of the rocker-arm.

The construction may now be completed by shifting the centre of the rocker-shaft T_1, so that the other arm shall have the proper motion with regard to nn', the centre-line of the valve-spindle. To make this construction, draw a line ll' parallel to nn', and at a distance equal to tq, taken from the triangle sqt or found by calculation from a proportion involving the side T_1q of the triangle eT_1q. With O as a centre and with a radius equal to OT_1 intersect the line ll' at T; this gives the position of the centre of the rocker-shaft. With T as a centre the arcs ba_0b' and dc_0d_1, on which the ends of the rocker-arms travel, can be drawn. At the same time that T_1 is swung to T, the point e_0 is swung to c_0, found by intersecting the arc dd_1 from O with a radius equal to Oe_0. From T draw Ta_0 perpendicular to nn'; then c_0Ta_0 is the position of the rocker when the valve is in mid-position, and aa' is equal to twice the lap. The position of the eccentric will be found by moving it through an angle equal to TOT_1', and in the same direction; thus when the crank is at OR_a' (the crank-end admission) the eccentric is at r_a'', found by making $r_a'Or_a'' = TOT_1$. The heavy black lines are drawn to show the eccentric, eccentric-rod, and rocker at the crank-end admission.

The true position of the eccentric is of importance only in setting the valve and need not be known exactly even then. The extreme positions of the eccentric-rod and of the rocker are to be found by trial, and from them the extreme positions b and b' of the head of the valve-rod or valve-spindle. Since this method of equalizing the cut-off introduces some irregularity, a complete study of the valve-motion, by aid of the valve-ellipse or otherwise, is desirable.

Piston-valve.—If the section of a plain slide-valve and its seat, as shown by Fig. 3, Pl. I, be supposed to revolve about an axis xx', there will be generated a piston-valve with its cylindrical seat. Such a valve is represented by Fig. 1, Pl. VII, which gives a section of the high-pressure cylinder and valve of the U. S. battle-ship *Massachusetts*. It will be seen that the outside shell of the cylinder, the lower cylinder-head, and the valve-chest form one casting, with feet attached for bolting to the engine-frame. A cylinder-liner is forced into and secured to the outer shell, with a space between to serve as a steam-jacket. The piston is of conical form, and the heads are shaped to correspond. Leakage is prevented by two packing-rings held in place by a junk-ring.

The piston-valve is in the shape of two pistons connected by a pipe or sleeve through which the valve-spindle passes. The valve-spindle is prolonged beyond the valve and attached to a small balancing piston which relieves the valve-gear of the weight of the valve and attached parts; the upper end of the balancing cylinder is connected with the exhaust. The valve-seat is formed by two short hollow cylinders, forced into the shell of the valve-chest. The space surrounding each half of the valve-seat is connected with and forms part of the passage leading to the cylinder. The ports are cut through the cylindrical valve-seat as shown.

Steam is supplied to the middle of the steam-chest, and is exhausted from the ends through pipes shown by dotted circles. This arrangement secures the advantages that the supply and exhaust steam are kept well separated so that heat cannot easily pass from one to the other, and the valve-rod stuffingbox is exposed to the exhaust steam only; such an arrangement is not advisable for a cylinder in which there may be a vacuum, since the leakage of air inward, past the stuffing-box, is more troublesome than the escape of steam. The laps controlling the admission and cut-off are by this arrangement placed inside; while the laps controlling the exhaust and compression are outside. To avoid confusion, it is advisable to distinguish them as *steam-lap* and *exhaust-lap;* the design of the valve-gear by the aid of valve-circles or otherwise may with this understanding be carried out as usual. It will be noticed that the top-end steam-lap is the larger; while the top end exhaust-lap is the smaller, and is here a negative lap or clearance. This arrangement is adopted to hasten the cut-off and compression on the down-stroke and delay them on the up-stroke, but is not carried far enough to produce complete equalization. The greater lead at the lower end helps to compensate for the shorter cut-off and the weight of the reciprocating parts.

Leakage past the valve is prevented by packing-rings, like those of the piston, which form the acting-edges of the valve. To prevent the valve-rings from springing into the ports, bridges are left as shown at A.

Double-ported Valve.—It is frequently difficult or impossible to get sufficient width of port for engines having a large diameter and short stroke, if the common plain slide-valve is to be used. Fig. 2, Pl. VII, shows a device, known as a double-ported valve, used in marine engineering to overcome this difficulty. Each passage leading to the cylinder has two ports, and two slide-valves, joined together and forming one casting, to control the flow of steam through those ports. The inner valve resembles the common slide-valve, except that there is a communication through the top between its exhaust-space e_0 and the exhaust-space e of the outer valve. The outer valve is elongated enough to leave a steam-space (a and a') to supply the inner valve; a bridge between e and a separates the exhaust of the outer valve from the steam-space of the inner valve, and is continued to the opening through the top of the inner valve. Fig. 3 gives, at the left hand, a transverse section through the exhaust space e, and at the right, through the steam-space a. The space a is drawn down toward the middle of the valve as shown, so that the valve may be made compact while providing sufficient area for the flow of steam.

Allen or Trick Valve.—Fig. 4, Pl. VII, shows a valve which is so made that a double admission of steam takes place at and near cut-off and admission. It is used with the link-motion and other gears which give a variable cut-off with the slide-valve, and is intended to remedy the defects due to the slow motion imparted to the valve at those points when the cut-off occurs early in the stroke.

Through the body of the valve there is a passage ss', and the valve-seat is cut away so that the distance from the outer edge of the passage to the edge of the valve-seat is equal to the outside lap of the valve. If the valve is displaced toward the right by the amount of the outside lap, the edge c of the valve is brought to the edge of the port a, and at the same time the edge of the passage s' is brought to the edge d of the valve-seat. Consequently there is a double admission of steam to the port a, one in the usual way past the edge c, and the other under the right-hand end of the valve, past the edge d of the valve-seat, and through the passage $s's$. As the valve opens wider, the passage ss' is liable to be shut off by traversing past the farther edge of the port a, but when that happens the supply past the edge c is abundant. Near cut-off the passage ss' is again opened at s and gives a double supply of steam till cut-off occurs by the simultaneous coincidence of c with the edge of the port a, and of d with the edge of the passage s'.

A modification of this form of valve, used on the Armington and Sims engine, is shown by Fig. 2, Pl. IX. The valve is a piston-valve, taking steam at the middle and exhausting at the ends, and differs from the valve shown by Fig. 1, Pl. VII, in that the stem connecting the two ends of the valve is hollow, and in that there is a passage from this interior channel through the valve-face. In the figure the valve is giving admission to the head end, and steam enters the cylinder directly from S at a, and also from S at b, through the hollow stem and thence through the passage in the valve-face.

Balanced Valves.—When the difference of pressure between the steam and exhaust pipes is large, the force exerted to hold a plain slide-valve against its seat is very large, and the friction of the valve on its seat is excessive. This consumes a needless part of the work developed by the engine, throws a severe duty on the valve-gear, and makes it difficult to maintain the acting-surfaces of the valve and its seat in good condition. Various methods of relieving valves from part or all of the steam-pressure on them have been devised, resulting in what are called balanced valves.

The piston-valve (shown by Fig. 1, Pl. VII) has no pressure on it to hold it against its seat, and is consequently perfectly balanced. It is very commonly used for the high-pressure and intermediate cylinders of triple-expansion marine engines, and on high-speed engines under the control of a shaft-governor. When well made, and provided with packing-rings, there is no more reason for leakage than exists with the piston of the engine. Small piston-valves are commonly made without packing-rings, and then depend on the fit in the valve-seat to prevent leakage. It is claimed that they do not leak when new, that the wear is insignificant, and that both valve and seat may readily be renewed when necessary. It is, however, probable that such piston-valves do frequently leak in common service.

The double-ported valve (Figs. 2 and 3, Pl. VII) and the Allen valve (Fig. 4, Pl. VII) have part of the pressure on the back relieved, and are known as balanced valves. The double-ported valve has a shallow cylindrical recess turned in its back. In this is a short cylinder or ring that is pressed by springs against a finished surface on the valve-chest cover. A bronze ring fastened to the valve and bearing against the vertical ring or cylinder is intended to prevent leakage. Communication is opened between the enclosed space and the exhaust, so that the leakage may not accumulate in this space and destroy the balancing of the valve. The unbalanced pressure of the steam on the unenclosed part of the valve gives enough pressure against the seat to prevent leakage. The Allen valve is commonly much longer than wide, and consequently a rectangular balancing-frame is used to exclude pressure from part of the top of the valve. Leakage into the enclosed space is allowed to flow directly into the exhaust-cavity through a small round passage, shown by dotted lines. All such devices are somewhat costly to make and troublesome to maintain in good condition, and if allowed to get out of condition are liable to a large loss from leakage directly into the exhaust.

Valve-setting.—A slide-valve is commonly set to give equal lead, or else equal cut-off. Sometimes the leads are made unequal, so as to partially equalize the cut-off; in this case the method of setting is like that used for equal lead, except that the lead at each end is made the amount determined on. If the cut-off is equalized by aid of a rocker or bell-crank lever, as shown on Plate VI, the valve is set to give equal lead.

As a preliminary to the setting of the valve, a method will be given for putting the engine-crank on the dead-centre.

To put an engine on the dead-centre.—In Fig. 1, Pl. VIII, let the circle $C_0CC_0'C'$ be the path of the crank, and let A_0A_0' be the stroke of the cross-head; while $abcd$ represents the edge of the fly-wheel or the face of the crank-disk, if the crank is so made. Set the engine with the cross-head near the middle of the stroke, and make reference-marks or take measurements so that it may be set again in the same position. Make a reference-mark on the circle $abcd$, and on some fixed object, at a and o. Turn the engine round till the cross-head again comes to A; the crank will then be at C', and the mark made at a will be found at c. Make another mark at a, and bisect the arcs abc and adc at b and d. It is apparent that the angular distance of b from a is equal to the angular distance of C' from C_0'; consequently the crank will be at the dead-point C_0', if the mark at b is brought opposite o. Also the crank will be at the dead-point C_0 when d is brought opposite o.

In this process, and during all the operations of valve-setting, the engine and the valve-gear should always be moved in the direction in which the engine is intended to run, so that the lost motion or back-lash may be taken up in the right way. Should the engine or the gear be moved too far at any time, then it should be turned back beyond the desired point, and brought up to that point with a motion in the right direction. Should the elasticity of the engine-belt interfere with the convenient and accurate setting of the engine, it may often be possible to place a stick of timber under a fly-wheel arm, block up one end and place a jack-screw under the other, and so force the engine to the desired setting and hold it at will; or some equivalent device may be used.

To set a valve with equal lead.—First method.—Set the engine on a dead-point and give the eccentric the proper angular advance, as near as may be; making it too much rather than too little. Adjust the length of the eccentric-rod or of the valve-spindle to give the valve the proper lead. Move the engine forward to the other dead-point, and measure the lead; if it is not right, then correct half the error by changing the length of the valve-spindle, and the other half by moving the eccentric. Repeat till the result is satisfactory.

If a valve-gear has a rocker, then the length of the valve-spindle should be such that the rocker may swing as designed; usually to an equal angle on each side of a perpendicular through its axis, to the centre-line of the eccentric-rod motion. In such case the eccentric-rod only should be changed in setting the valve; a small change of the valve-spindle may be allowed.

Second method.—A valve that has harmonic motion will give the same maximum port-opening when set with equal lead. Such a valve may be set for equal lead by the following method. Valves which do not have harmonic motion cannot be so set; as examples may be quoted a slide-valve having equal lead and with the cut-off equalized by aid of a rocker or bell-crank lever, and a valve controlled by a link-motion or radial valve-gear; the two last forms will be described in future chapters:

Loosen the eccentric on the shaft, and turn it around till it gives the maximum port-opening first at one end and then on the other. If the maximum port-openings are not equal, make them so by changing the length of the valve-spindle by half the difference; this operation adjusts the length of the valve-spindle. When that adjustment is complete, set the engine on a dead-point and give the valve the proper lead by turning the eccentric on the shaft; this adjusts the angular advance. This method is convenient when it is difficult to turn the engine.

To set a valve for equal cut-off.—With the crank on the head-end dead-point, give the eccentric the proper angular advance, and give the valve the proper lead. Move the engine forward till cut-off occurs, and measure the displacement of the cross-head from the beginning of the stroke. Move the engine forward, again, till cut-off occurs on the return stroke, and measure the displacement of the cross-head from the crank end of the stroke. Should the cut-off be earlier at the head end than at the crank end, the valve-spindle is too long; and conversely it is too short if the crank-end or return-stroke cut-off is the earlier. In either case, change the length of the valve-spindle by an amount which it is estimated will correct the inequality; it may be convenient to draw a valve-diagram to aid in making an estimate for a large engine. Set the engine again on the head-end dead-point, and adjust the lead by moving the eccentric. Try the cut-off again, and repeat till the result is satisfactory.

It is apparent that a valve that is designed for equal cut-off will be properly set if the leads are made what the design gives for them. When such information is at hand, the process of setting the valve will be the same as the first method except that the lead at each end is to be made the proper amount, with the addition that the displacement of the cross-head is to be determined for each stroke, and the adjustment is to be completed by the method just given.

To set a valve with the steam-chest cover on.—Figs. 2 and 3, Pl. VIII, show an arrangement by aid of which the valve may be set with the steam-chest cover on, and, if convenient, with steam applied to the engine. In preparation the valve is set to give the proper lead at each end, and a centre-punch mark is made on the valve-spindle outside of the stuffing-box. A pair of trams are made of heavy wire, and adjusted so that they shall reach from the mark on the valve-spindle to a mark on the steam-chest, one at one dead-point and one at the other. In setting the engine, the first method for equal lead is to be used, with the difference that the valve is set to give the lead, when the engine is on a dead-point, by aid of the trams. A variation of this method may be used, for which one tram is required and two centre-punch marks are made on the valve-spindle. The method is convenient for locomotives, and it is customary to provide trams for this purpose in the tool-chest of the locomotive, so that the valves may be set, if necessary, on the road.

It should be noted that the method of setting the valve with equal lead or equal cut-off insures that the action of the valve shall be what is desired when opening or closing. Any error of design due to the neglect of the angularity of the eccentric-rod is therefore transferred to some other part of the motion of the valve, namely, to a place where the valve is open or closed, and any irregularity of motion is then of little consequence.

CHAPTER II.

ADJUSTABLE ECCENTRICS.

Reversing with Loose Eccentric.—The device shown by Fig. 4, Pl. VIII, was used for reversing some of the earliest locomotive and marine engines; it is to-day used for reversing small and unimportant engines, and with some modifications to secure positive action, is used on engines of considerable power. As shown, the crank is at C, and the eccentric has its centre at E, so that the engine will run in right-handed rotation as shown by the arrow. The eccentric is loose on the shaft and has a pin at B, which engages with the end of a circular slot in a disk back of the eccentric, so that the eccentric is driven by the disk. To reverse the engine, it is stopped, and the eccentric or the engine is turned till the pin B engages with the other end of the slot at B'.

The valve-circles for forward and for backward motion are drawn at OP and OP', and the lap-circle is nn_1n_2; the cut-off occurs at OR, on the forward stroke when running right-handed, and at OR', on the forward stroke when running reversed. The valve-circles for the return stroke are omitted to avoid confusion.

Shifting Eccentric with Variable Lead.—A shifting eccentric, like that shown by Fig. 1, Pl. X, may be used for reversing an engine, and it possesses also the property of giving a variable cut-off. The eccentric is swung on a pivot S, a point on the centre-line of the eccentric and eccentric rod motion, and is slotted to clear the shaft; the angle OSO' is made equal to ESE', so that the centre of the eccentric may be brought to the point E' when the engine is reversed. Let the lap of the valve be equal to Ob; then the displacement of the valve when the engine is on a dead-point is Oa, found by drawing the vertical EaE', and the lead is ba.

In Fig. 2 the valve-circle OP represents the valve-motion when the eccentric-centre is at E. The cut-off occurs at the crank-position OR, or at the piston-displacement xa, assuming harmonic motion; with crank and connecting-rod the piston-displacement will be longer at the forward-stroke cut-off, and shorter at the return-stroke cut-off, but in such case xa is nearly the mean for the two strokes. When reversed the motion of the valve will be represented by the dotted valve-circle OP'.

Suppose now that the eccentric-centre is shifted to E_1, Fig. 1; the angular advance is YOE_1, and the eccentricity is OE_1. The valve-circle OP_1 will represent the motion of the valve; it has the angle YOP_1 equal to the angular advance, and the diameter OP_1 equal to the eccentricity when the eccentric-centre is at E_1, Fig. 1. The point P_1 is evidently on an arc of a circle having its centre on XOX' produced, and drawn with a radius equal to ES, Fig. 1. In like manner the valve-circles OP_2 and OP_0 represent the motion given to the valve when the centre of the eccentric is at E_2 and at E_0, respectively. The piston-position at cut-off for the valve-circle OP_1 is a_1; the cut-off for the valve-circle OP_2 is at a_2, and for the valve-circle OP_0 is at a_0. Thus it appears that the cut-off may be made to vary from the piston-displacement xa to the piston-displacement xa_0; that is, from $\frac{7}{8}$ to $\frac{1}{8}$ of the stroke. All the other events of the stroke, namely, compression, release and admission, vary at the same time as the cut-off, and in a similar manner, though to a less degree. Of these other events, the admission varies the least; an examination of the figure will show that the lead-angle increases from about 2° to about 40°. The lead increases from ba to bE_0, Fig. 1, and the increase is as clearly shown by Fig. 2. A circle representing the inside lap would, if drawn, show the change of release and compression;

it is omitted to avoid further complexity. Such a circle would show that the compression varies less than the cut-off, and that the release varies more than the admission. It is apparent that the increase of the lead depends on the radius SE (Fig. 1) of the arc on which the eccentric-centre moves, and may be diminished by moving the pivot S away from the axis of the shaft.

The gear is said to be in full-gear forward when the eccentric is at OE, Fig. 1, and at full-gear backing when the eccentric is at OE'. When the eccentric is at OE_0 the gear is said to be at mid-gear; intermediate positions may be called grades. An examination of Fig. 2 will show that, since the centre of the valve-circle OP_0, is on the axis XOX', the crank-angle at admission is equal to the crank-angle at cut-off, and this, with other considerations, will indicate that the mid-gear position of the eccentric does not give a proper motion of the valve for either forward or backing motion of the engine.

Shaft-governor.—At the present time the shifting eccentric is widely used on high-speed engines, which have the valve under the control of a shaft-governor. Fig. 1, Pl. XI, shows the valve-gear of the Straight Line engine. The line XX' is the centre-line of the piston, crank and connecting-rod; xx' is the line of the valve-spindle. The cut-off is equalized for one grade by aid of a rocker, and is not far from equal at other grades. The eccentric is pivoted at S, so that OS is the centre-line of the eccentric and eccentric-rod motion. The rocker oscillates on a centre at T, and the arms Ta and TC are connected, respectively, with the valve-spindle xx' and the eccentric-rod CE. The governor consists of the weight W and the spring LQ. The governor-lever WNM is pivoted at N, and is connected by the link MVL to the eccentric casting at V, and to the five-leaved spring QL at L. When the engine is at rest the weight W lies against the fly-wheel boss as shown, and the eccentric is at full-gear. When the engine comes up to speed, the centrifugal force acting on W enables it to compress the spring QL, and at the same time to move the eccentric toward mid-gear, and thus the governor adjusts the cut-off, and consequently the steam-supply, to the load. Since the engine is never reversed, the eccentric is slotted to clear the shaft only far enough to bring its centre to mid-gear. Fig. 2 shows the valve-diagrams for this gear drawn to an enlarged scale.

In order that a shaft-governor may be able to control the valve of an engine, and maintain the speed nearly uniform, without being of excessive size, the valve must be nicely balanced and must move very freely. The Straight Line engine has a flat valve that moves between its seat and a cover-plate with enough clearance to avoid friction, but not enough to allow of leakage, just as a piston-valve moves in its cylindrical seat. A device similar in effect to the passage through the body of the Allen valve gives a double admission at and near cut-off and admission. The piston-valve, usually without packing-rings, is commonly used in connection with the fly-wheel governor.

The action of the reciprocating parts of a high-speed engine is of great importance. A considerable part of the work of the steam is expended in imparting motion to the reciprocating parts during the first half of the stroke, and this stored energy is restored during the second half of the stroke as the reciprocating parts come to rest. In order that they may come to rest quietly at the end of the stroke the piston should be cushioned by compression. Now a valve that gives a variable cut-off and a variable compression is likely to have too little compression at full-gear and too much at a short cut-off. An engine with a large clearance will suffer less from this difficulty than one with a small clearance; consequently the clearance of high-speed engines with shaft-governors is often made large, but a large clearance is not conducive to economy in the use of steam. Now, lead acts, as does compression, to stop the reciprocating parts and to fill the cylinder with steam, so that in general the more compression an engine has the less lead it will need. But it has just been seen that the shifting eccentric shown on Plate X has an increasing lead toward mid-gear, that is, at the time when it is least needed on a stationary engine. Had the pivot S been placed on the other side of the shaft, then the lead would have decreased toward midgear. Fig. 1, Pl. XII, gives the valve-diagrams for such an eccentric. Some forms of the Straight Line engine have their valve-gears so arranged in order that the decreasing lead toward mid-gear may partially compensate for the increasing compression.

Shifting Eccentric with Constant Lead.—Fig. 2, Pl. XII, shows an eccentric that has a motion square across the shaft, thus carrying the centre of the eccentric on the straight line EE_0E' from full-gear forward to full-gear backing. It is at once apparent that the lead is constant. Fig. 3 gives the valve-diagrams for the full-gear, mid-gear, and two intermediate grades. Though the lead is constant, the lead-angle is not so; a comparison with Fig. 2, Pl. X, will show that the variation is not so much as that of the lead-angle for an eccentric with increasing lead, but it is more than for an eccentric with decreasing lead, as may be seen by a comparison with Fig. 1, Pl. XII.

A shifting eccentric with constant lead must be slotted to clear the shaft; the line OO' is made equal to EE', in order that E may pass to EE' when the engine is reversed. If the engine runs always in one direction, the slot, from centre to centre, may be only as long as EE_0.

The shaft-governor of the Armington and Sims engine, shown by Fig. 1, Pl. IX, is equivalent to a shifting eccentric with constant lead. The eccentric E is on a casting, ab, loose on the shaft; the eccentric E_1 is loose on the eccentric E. The weights W and V compress the springs N and M, and move out toward the rim of the wheel as the engine comes up to speed. The links ac and bd take hold of the eccentric-casting ab, and the link st takes hold of the eccentric E_1. As the weights move out the linkage $cabd$ takes a new position, such as $c'a'b'd'$, turning the eccentric E toward the left; at the same time the link st takes the position $s't'$, turning the eccentric E_1 toward the right. The proportions of the mechanism are so chosen that the centre, e, of the outer eccentric, E_1, moves straight across the shaft to a new position, e', and the cut-off is shortened. It will be seen that the effect is the same as that of the shaft-governor of the Straight Line engine, except that the lead is constant.

CHAPTER III.

LINK-MOTION.

THE valves of locomotives, marine engines, and other reversing engines are commonly controlled by a mechanism called a link-motion; this mechanism has also the property of giving a variable cut-off. The mechanism consists essentially of two eccentrics, one for full-gear forward and one for full-gear backing, together with the eccentric-rods and the *link*. The eccentric-rods are attached to the link, at or near the ends, and the link is slotted or otherwise arranged to receive a block on the end of the valve-spindle, on a radius-rod, or the end of a rocker, as the case may be. The link-motion takes two forms; in one, known as the Stephenson or shifting link, the link is moved on the block to reverse the engine or vary the cut-off; in the other, known as the Gooch or stationary link, the block is moved in the link to accomplish the same object.

Stephenson Link.—The usual form of link-motion for American locomotives is shown by Figs. 1 and 2, Pl. XIII. The valve is moved through a rocker so that the eccentrics follow the crank;. thus, the centre of the crank is at C, and the go-ahead eccentric has its centre at E, while the backing eccentric centre is at E'. The link-pins P and P', to which the eccentric-rods are attached, are set back from the link-arc, and the link may move over the link-block B so far as to bring the centre of the block opposite the centre of the link-pin, as shown by Fig. 1; in which position the motion of the valve is controlled almost entirely by the eccentric E, and has essentially the motion of a plain slide-valve. The link is suspended by a link or hanger, nm_0, from a reverse-shaft centred at S; the hanger takes hold of the saddle-pin m_0 on a plate that is commonly at the middle of the link.

A locomotive has two cylinders, with pistons acting on two cranks set at a right angle, and thus has two engines each of which must be provided with its own link-motion. Both links are suspended from one reverse-shaft, which has an arm SR from which a rod runs to a reverse-lever conveniently located in the engineer's cab. The reverse-lever moves over a notched arc, and by aid of a latch engaging with the notches the link may be set and secured in any desired position.

Fig. 2, Pl. XIII, gives an end elevation of the link, the hanger, and one arm of the rocker carrying the link-block B.

English locomotives commonly have the link act directly on the valve-spindle, without the intervention of a rocker. In such case the link-pins should be placed on the link-arc, as shown by Fig. 3, Pl. XIII, and consequently the link-block cannot be opposite one of the link-pins and cannot receive the full motion of the eccentric. Consequently the eccentricity, and with it all the dimensions of the link-motion, must be larger to give proper motion to the valve.

The link-motion for the high-pressure cylinder of one of the engines of the U. S. S. *Maine* is shown by the figures on Plate XIV. The parts are lettered as on Plate XIII; thus E and E' are the centres of the eccentrics, and P and P' are the link-pins, which in this case are on the link-arc; S is the reverse-shaft, and NP is the drag-link or bridle which takes hold of the go-ahead link-pin. The link, which is known as the Scotch or side-bar link, is shown in plan by Fig. 2. The link-block is between the side-bars, and is pivoted directly on the end of the valve-spindle; thus the link can be set so that the axis of the link-pin coincides with that of the link-block pivot, and the full motion of the eccentric can be given to the valve. The head of the valve-spindle is guided by cast-iron jaws, as shown in Fig. 1. The end of the reverse-arm is slotted and provided with a sliding-block, screw, and hand-wheel as shown, so that the cut-off may be adjusted in a manner to be described later.

Open Rods and Crossed Rods.—If the eccentric-rods of a link-motion are connected as shown in Fig. 1, Pl. XV, the rods are said to be open; on the other hand, the rods are said to be crossed when connected as shown by Fig. 2. In both figures the crank is on the crank-end dead-point, and the valve-gear has no rocker. A link-motion with a rocker is said to have open rods when the eccentric-rods are connected as shown by Fig. 3; aTb is the rocker, and bc the valve-spindle; the crank is on the head-end dead-point. In all these figures the link-pins are on the link-arc. Since half a revolution will apparently cross the rods for Figs. 1 and 3, while it will apparently open the rods for Fig. 2, the nomenclature seems to be unfortunate. There is, however, a real difference in the methods of the connection of the rods, and that difference has an important influence on the action of the valve, for open rods give an increasing lead from full toward mid gear, while crossed rods give a decreasing lead from full toward mid gear. In Fig. 1, the full lines Ep_0, $E'p_0'$, and the arc $p_0 b_0'$, show the eccentric-rods and the arc of the link at mid-gear, while the thin lines Ec, $E'p'$, and the arc cp', show them at full-gear forward. Since the valve and valve-rod have the same motion as the link-block, it will be sufficient to trace the motion of the latter. At full-gear the link-block will be at c, found by intersecting the line of centres with E as a centre, and with a radius equal to the length of the eccentric-rod. The eccentric-pin p' is located by drawing arcs from E' and c, with the lengths of the eccentric-rod and the length of the link as radii. At mid-gear the link-block is ac_0; the points p_0 and p_0' are at a distance from the centre-line OX, equal to half the chord of the link-arc, and the link is erect. The increase of lead from full-gear to mid-gear is apparent from the diagram. A similar construction in Fig.

2 shows the decrease of lead from full-gear toward mid-gear for crossed rods. In the figure the decrease is greater than the full-gear lead, so that the valve is shut at the dead-point when the link is at mid-gear.

Long and Short Rods.—The variation of lead from full-gear toward mid-gear is due to the curvature of the link-arc, and is more pronounced for a link with short radius than for one with long radius; now the radius of the link-arc is usually equal to the length of the eccentric-rod, hence the variation is more for short than for long rods. In Fig. 1, Pl. XV, it is apparent that $c''c_0''$ is greater than cc_0; a similar construction will show that the decrease of the lead from full-gear to mid-gear for crossed rods is more marked for short than for long rods.

Radius of the Link-arc.—An analytical discussion of the link-motion shows that the radius of the link-arc should be equal to the length of the eccentric-rod; if the link-pins are on the link-arc, then the radius should be the distance from the centre of the eccentric to the link-pin; but if the pins are back of the arc, the radius is the distance from the centre of the eccentric to the link-arc, i.e. the length of the rod plus the distance the pins are back of the arc. The same discussion establishes also the fact that open rods give increasing lead, and that crossed rods give decreasing lead, from full-gear toward mid-gear; but the demonstrations given are believed to be useful, and a similar demonstration will be given of the proper radius for the arc.

In Fig. 1, Pl. XV, the link-block is at c at full-gear when the crank is on the crank-end dead-point; when the crank is on the head-end dead-point a similar construction will give for the position of the link-block the point c'. The point o, half way between c and c', corresponds to the mid-position of the valve, and from o the lap on, on' may be laid off on each side, giving $nc = n'c'$ for the lead. At mid-gear the head-end lead is nc_0, and a similar construction for the head-end dead-point will give $n'c_0' = nc_0$ for the head-end mid-gear lead. If now a diagram is drawn for some intermediate gear of the link, it will be found that the lead is the same at the two ends, and that it is intermediate between nc and nc_0. Fig. 1 is drawn with the radius of the link-arc equal to the length of the eccentric-rod, and any diagram drawn with dimensions chosen at random will in like manner show equal leads under like conditions. Constructions for a link with crossed rods will also show equal leads if the radius of the link-arc is equal to the length of the eccentric-rod. Consequently it may be inferred that the one requirement for equal leads is that given, i.e., that the radius of the link-arc shall be equal to the length of the eccentric-rods. A natural inference is that any other radius for the link-arc will give unequal leads for some grades of the link, and such will be found to be the case if constructions are made.

Analytical Discussion.—The best idea of the nature of the motion given by a link to the valve is obtained from an analytical discussion due to Zeuner, taken, with slight variations, from his *Treatise on Valve-gears.*

On Plate XVI the Figures 1 and 2 are drawn to represent link-motions with open and with crossed rods. The diagrams in thin lines give the positions of the parts when the crank is on the crank-end dead-point; and the diagrams in heavy lines show the positions when the crank has moved through the angle θ.

The eccentricity for each eccentric is r, and the angular advance is δ. The link-pins are on the link-arc, the length of the eccentric-rods is l, and the radius of the link-arc is ρ which may or may not be equal to l. The length of half the link-arc is c, and the displacement of the link-block from the middle of the link-arc is d. In the discussion it is assumed that the link is supported and guided by the link-block so that the point n remains on the line XX'. It is also assumed that the chord joining the link-pins is equal to the length of the link-arc between those pins, and that in like manner the displacement of the link-block from the middle of the link may be measured indifferently either on the link-arc or on the chord. The error of this assumption may be estimated as follows: A common proportion is $r = \frac{1}{2}c = \frac{1}{12}l$, or $c = \frac{1}{6}l$, so that the arc subtends an angle of about 10°, and for that angle the arc is 0.1745 of the radius and the chord is 0.1736 of the radius, and the error is a little more than half of one per cent.

The distance from the centre of the driver-axle to the middle of the valve, in either Fig. 1 or Fig. 2, Pl. XVI, is

$$Ob = Om + mn + nb = Op - mp + mn + nb, \quad . \quad (7)$$

in which the length of the valve-spindle nb may be replaced by s, and the value of the other terms may be conveniently determined as follows:

First, the term mp is determined by the equation

$$mp = mP \sin mPp = (c - d) \sin \alpha. \quad . \quad . \quad . \quad (8)$$

Now

$$\sin \alpha = \frac{pp'}{PP'} = \frac{Op - Op'}{2c}. \quad . \quad . \quad . \quad . \quad (9)$$

From Fig. 1, Pl. XVI,

$$Op = Oe + ep = Oe + \{\overline{EP}^2 - (Pp - Ee)^2\}^{\frac{1}{2}}; \quad . \quad (10)$$

and from Fig. 2,

$$Op = Oe + ep = Oe + \{\overline{EP}^2 - (Pp + Ee)^2\}^{\frac{1}{2}}. \quad . \quad (11)$$

LINK-MOTION. 45

In either figure

$$Oe = r \sin(\theta + \delta), \quad Ee = r \cos(\theta + \delta), \quad EP = l,$$

and

$$Pp = mP \cos mPp = (c - d) \cos \alpha;$$

these values substituted in equations (10) and (11) give

$$Op = r \sin(\theta + \delta) + \{l^2 - [(c - a)\cos\alpha \mp r \cos(\theta + \delta)]^2\}^{\frac{1}{2}}; \quad (12)$$

the upper sign being taken for open and the lower sign for crossed rods. Expanding the term with a fractional exponent by the binomial theorem, and rearranging terms with the higher powers of l in the denominator, gives

$$Op = r \sin(\theta + \delta) + l - \frac{(c - d)^2 \cos^2\alpha}{2l}$$
$$\pm \frac{(c - d)r \cos(\theta + \delta)\cos\alpha}{l} + \frac{r^2 \cos^2(\theta + \delta)}{2l}.$$

Now the terms containing $\cos \alpha$ in the numerator have l in the denominator, and are small compared with $r \sin(\theta + \delta)$, while α is not more than 30°, for which the cosine is 0.866; consequently we may replace unity for $\cos \alpha$ without much error. With that change and some expansion,

$$Op = r \sin(\theta + \delta) + l - \frac{c^2}{2l} + \frac{cd}{l} - \frac{d^2}{2l}$$
$$\pm \frac{(c - d)r \cos(\theta + \delta)}{l} - \frac{r^2 \cos^2(\delta + \theta)}{2l}. \quad (13)$$

Similarly Fig. 1 and Fig. 2 give respectively equations (14) and (15):

$$Op' = e'p' - Oe' = \{\overline{E'P'}^2 - (P'p' - E'e')^2\}^{\frac{1}{2}} - Oe'; \quad (14)$$

$$Op' = e'p' - Oe' = \{\overline{E'P'}^2 - (P'p' + E'e')^2\}^{\frac{1}{2}} - Oe'. \quad (15)$$

In either figure

$$Oe' = r\sin(\theta - \delta), \quad E'e' = r\cos(\theta - \delta), \quad E'P' = l,$$

and

$$P'p' = mP'\cos mP'p' = (c+d)\cos \alpha;$$

these terms substituted in equations (14) and (15) give

$$Op' = -r\sin(\theta - \delta) + \{l^2 - [(c+d)\cos\alpha \mp r\cos(\theta-\delta)]^2\}^{\frac{1}{2}}; \quad (16)$$

the upper sign being taken for open and the lower for crossed rods. As in the previous work, the cos α may be replaced by unity; expanding by the binomial theorem, and rejecting terms with the higher powers of l in the denominator, gives

$$Op' = -r\sin(\theta - \delta) + l - \frac{(c+d)^2}{2l}$$
$$\pm \frac{(c+d)r\cos(\theta - \delta)}{l} - \frac{r^2\cos^2(\theta - \delta)}{2l};$$

$$\therefore Op' = -r\sin(\theta - \delta) + l - \frac{c^2}{2l} - \frac{cd}{l} - \frac{d^2}{2l}$$
$$\pm \frac{(c+d)r\cos(\theta - \delta)}{l} - \frac{r^2\cos^2(\theta - \delta)}{2l}. \quad (17)$$

Substituting in equation (9) the values for Op and Op',

$$\sin \alpha = \frac{r \sin (\theta + \delta) + r \sin (\theta - \delta)}{2c} + \frac{2cd}{2cl}$$
$$\pm \frac{(c-d)r \cos (\theta + \delta) - (c+d)r \cos (\theta - \delta)}{2cl}$$
$$- \frac{r^2 \cos^2 (\theta + \delta) - r^2 \cos^2 (\theta - \delta)}{4cl};$$

$$\therefore \sin \alpha = \frac{r}{c} \cos \delta \sin \theta \mp \frac{r}{l} \sin \delta \sin \theta \mp \frac{dr}{cl} \cos \delta \cos \theta$$
$$+ \frac{d}{l} - \frac{r^2}{4cl}[\cos^2 (\theta + \delta) - \cos^2 (\theta - \delta)]. \quad (18)$$

To find the value of the term mn, which in Fig. 3, Pl. XVI, is seen to be nearly equal to $m_o i$, we have

$$mn = m_o i = m_o n_o - i n_o;$$
$$\therefore mn = \frac{\overline{Pm_o}^2}{m_o T} - \frac{\overline{ni}^2}{iT} = \frac{c^2}{2\rho} - \frac{d^2}{2\rho} \text{ (nearly).} \quad (19)$$

Substituting in equation (7) the values of the several terms,

$$Ob = r \sin \delta \cos \theta + r \cos \delta \sin \theta + l - \frac{c^2}{2l} + \frac{cd}{l} - \frac{d^2}{2l}$$
$$\pm \frac{r(c-d)}{l}(\cos \theta \cos \delta - \sin \theta \sin \delta) - \frac{r^2 \cos^2 (\delta + \theta)}{2l}$$
$$- (c-d) \left\{ \frac{r}{c} \cos \delta \sin \theta \mp \frac{r}{l} \sin \delta \sin \theta \mp \frac{dr}{cl} \cos \delta \cos \theta \right.$$
$$\left. + \frac{d}{l} - \frac{r^2}{4cl}[\cos^2 (\theta + \delta) - \cos^2 (\theta - \delta)] \right\}$$
$$+ \frac{c^2}{2\rho} - \frac{d^2}{2\rho} + s;$$

$$Ob = r[\sin \delta \pm \frac{c-d}{l} \cos \delta \pm (c-d)\frac{d}{cl} \cos \delta] \cos \theta$$

$$+ r[\cos \delta \mp \frac{c-d}{l} \sin \delta - \frac{c-d}{c} \cos \delta \pm \frac{c-d}{l} \sin \delta] \sin \theta$$

$$- \frac{r^2}{4cl}\left\{ 2c \cos^2(\delta + \theta) - (c-d)[\cos^2(\theta+\delta) - \cos^2(\theta-\delta)] \right\}$$

$$+ \frac{-c^2\rho + 2cd\rho - d^2\rho - 2(c-d)d\rho + c^2l - d^2l}{2l\rho} + l + s;$$

$$\therefore\ Ob = r\left(\sin \delta \pm \frac{c^2-d^2}{cl} \cos \delta\right)\cos \theta + r\frac{d}{c} \cos \delta \sin \theta$$

$$- \frac{r^2}{4cl}\left[(c+d) \cos^2(\delta + \theta) + (c-d) \cos^2(\theta - \delta)\right]$$

$$+ (c^2 - d^2)\frac{l-\rho}{2l\rho} + l + s. \quad \ldots\ldots\ldots\ (20)$$

The third term has its greatest value when d is equal to c, and it is then equal to

$$\frac{r^2 \cos^2(\theta + \delta)}{2l},$$

which is the term that appears in equation (5) for the plain slide-valve. In the discussion of the plain slide-valve this term was neglected, and consequently it may be neglected here with equal propriety. Equation (20) may therefore be written

$$Ob = r\left(\sin \delta \pm \frac{c^2-d^2}{cl} \cos \delta\right) \cos \theta + r\frac{d}{c} \cos \delta \sin \theta$$

$$+ (c^2 - d^2)\frac{l-\rho}{2l\rho} + l + s. \quad . \quad (21)$$

If the engine is on the crank-end dead-point, then θ is zero; and it is 180° at the head-end dead-point. The special values of the crank-angle give

$$Ob' = r\left(\sin\delta \pm \frac{c^2 - d^2}{cl}\cos\delta\right) + (c^2 - d^2)\frac{l-\rho}{2l\rho} + l + s; \quad (22)$$

$$Ob'' = -r\left(\sin\delta \pm \frac{(c^2 - d^2)}{cl}\cos\delta\right) + (c^2 - d^2)\frac{l-\rho}{2l\rho} + l + s. \quad (23,$$

The mid-position of the valve should be midway between b' and b'', Figs. 1 and 2, Pl. XVI; half the sum of Ob' and Ob'' is

$$Oo = \frac{Ob' + Ob''}{2} = (c^2 - d^2)\frac{l-\rho}{2l\rho} + l + s; \quad . \quad (24)$$

in which the only variable is the term containing ρ, the radius of curvature of the link-arc. If ρ be made equal to l, then this term disappears, leaving

$$Oo = l + s. \quad \ldots \ldots \quad (25)$$

With equal laps, the necessary and sufficient condition for equal leads, at all grades, is that the radius of the link-arc shall be equal to the length of the eccentric-rod.

Applying this condition to equation (21) gives

$$Ob = r\left(\sin\delta \pm \frac{c^2 - d^2}{cl}\cos\delta\right)\cos\theta + r\frac{d}{c}\cos\delta\sin\theta + l + s. \quad (26)$$

50   VALVE-GEARS FOR STEAM-ENGINES.

The displacement of the valve from mid-position is

$$e = Ob - Oo;$$

$$\therefore \ e = r\left(\sin \delta \pm \frac{c^2 - d^2}{cl} \cos \delta\right) \cos \theta + r\frac{d}{c} \cos \delta \sin \theta. \quad (27)$$

General Equation for Valve-motion.—The equation (6) gives for the displacement of a plain slide-valve moved by an eccentric,

$$e = r \sin (\delta + \theta);$$

expanding the parenthesis,

$$e = r \cos \delta \sin \theta + r \sin \delta \cos \theta; \quad \ldots \quad (28)$$

which may be written

$$e = A \cos \theta + B \sin \theta, \quad \ldots \ldots \quad (29)$$

since r and δ are constant for a given slide-valve gear.

It has been shown by the aid of Fig. 7, Pl. II, that the motion of a plain slide-valve may be represented by a valve-circle, and a comparison of that figure with the equation (29) will show that the constants in the equation are the coördinates of the end P of the diameter of the valve-circle. Thus

$$Oq = r \sin \delta = A; \quad \ldots \ldots \quad (31)$$

$$Pq = r \cos \delta = B. \quad \ldots \ldots \quad (32)$$

It may be concluded that any valve which has its displacement represented by an equation of the same form as equation (29) has a harmonic motion and may have its motion represented by a valve-circle.

Zeuner's Diagram.—A comparison of equation (27) with equation (29) shows that a valve controlled by a Stephenson link-motion has a harmonic motion, and that its displacements from mid-position, at any grade of the link, may be represented by a valve-circle, having for the coördinates of the end of the diameter

$$A = r\left(\sin \pm \frac{c^2 - d^2}{cl} \cos \delta\right); \quad \ldots \quad (33)$$

$$B = r\frac{d}{c}\cos \delta. \quad \ldots \ldots \ldots (34)$$

At full-gear $d = c$, which applied to equations (33) and (34) will reduce them to equations (31) and (32). Thus the valve-diagram for a link-motion at full gear is identical with the diagram for a plain slide-valve under the control of an eccentric having the same eccentricity and angular advance as one of the eccentrics of the link-motion; which coincides with our conceptions of the link-motion, derived from the drawings on Plates XIII and XIV.

In Fig. 4, Pl. XVI, and Fig. 1, Pl. XVII, Oq and qP are made equal to the coördinates of the end of the valve-circle diameter at full-gear, when $d = c$; i.e.,

$$Oq = r \sin \delta = A, \quad qP = r \cos \delta = B.$$

At mid-gear d becomes zero, and the coördinates the end of the diameter of the valve-circle become

$$A_0 = r \sin \delta \pm \frac{c}{l} \cos \delta; \quad \ldots \ldots (35)$$

$$B_0 = 0. \quad \ldots \ldots \ldots \ldots (36)$$

The upper sign is taken for open rods, and the lower sign for crossed rods. Fig. 4, Pl. XVI, corresponds with the first case, and Fig. 1, Pl. XVII, with the second case. The ends of the diameter of valve-circles for intermediate grades of the link may be found by assuming values for d and calculating the coördinates by aid of equations (33) and (34), if desired; but an inspection of those equations shows that they give the coördinates of a parabola having its vertex on the axis XX'. Two points, P and P_0, are already located and an arc of the parabola may be passed through them by the ordinary geometrical construction; or, since the arc is quite flat, there may be substituted for it the arc of a circle having its centre on the axis XX'. The centres of the valve-circles have for their coördinates, $\tfrac{1}{2}A$ and $\tfrac{1}{2}B$, and consequently lie on another parabola with its vertex on XX'; an arc of a circle centred on XX' may be substituted for the arc of the parabola, and will be more convenient to draw since its radius is half the radius of the circular arc substituted for the parabola through P and P_0.

Fig. 4, Pl. XVI, and Fig. 1, Pl. XVII, exhibit the variation of lead which was pointed out in Figs. 1 and 2, Pl. XV. The same thing is evident from an inspection of equation (33), taking the upper sign for open and the lower sign for crossed rods. For convenience the fact is stated as follows:

A Stephenson link-motion with open rods gives increasing lead from full-gear toward mid-gear; with crossed rods it gives decreasing lead from full-gear toward mid-gear.

Valve-circles showing the motion of the valve at intermediate grades of the link are drawn at OP_1 and OP_2, on Fig. 4, Pl. XVI, and on Fig. 1, Pl. XVII. On both figures the lap-circles are $nn'n''$, showing cut-off at the crank-positions OR, OR_1, OR_r, and OR_0; neglecting the influence of the connecting-rod, the corresponding piston-displacements are xa, xa_1, xa_1, and xa_0. An inspection of the figure will show that the lead-angle increases as the cut-off is shortened, accompanied by an earlier admission. If an inside lap-circle were drawn, it would show that an early cut-off is accompanied by an early release and a large compression. A comparison of these diagrams with the valve-diagrams shown by Fig. 2, Pl. X, and Fig. 1, Pl. XII, will show that a Stephenson link-motion is equivalent to a shifting eccentric with variable lead.

Gooch Link.—Fig. 1, Pl. XVIII, shows the Gooch or stationary link, as applied to locomotive engines. E and E' are the forward and backing eccentrics, from which the eccentric rods lead to the link-pins P and P'. The link is suspended by the link mn, from a fixed pivot n, and has its convex side turned toward the axle O. The link-block B is carried by a radius-rod BD, which is connected directly to the head of the valve-spindle at D. By means of a reverse-arm ST and hanger TU, the engineer may place the link-block B opposite the link-pin P for full-gear forward, opposite the link-pin P' for full-gear backing, or at any intermediate position. The action of this link-motion is therefore equivalent to that of the Stephenson link-motion. The details of the mechanism are varied somewhat by different makers. In the figure the link is suspended from a saddle-pin on or near the chord joining the ends of the link-arc; and for this purpose a plate or bridge similar to that shown by Fig. 2, Pl. XIII, is employed which permits the passage of the link-block. The saddle-pin is sometimes placed behind the link-arc toward O, so as to avoid the use of a plate or bridge. The link-pins are placed behind the link-arc to allow the link-block to be brought opposite the link-pins. They may be placed on the link-arc, using a link like that shown by Fig. 3, Pl. XIII, but turned so that the convex side is toward the axle O; in which case the full action of the eccentric cannot be given to the valve. Sometimes a box-link, shown by Fig. 3, Pl. XVIII, is used, and then the saddle-pin and link-pins may be placed in any desired positions without interfering with the link-block; this device is equivalent to the side-bar link shown on Plate XIV.

Open and Crossed Rods.—As was found to be the case with the Stephenson link, the rods of the Gooch link may be open or crossed. Fig. 1, Pl. XVIII, has open rods, and Fig. 2, Pl. XVII, has crossed rods.

Radius of the Link-arc.—The common and proper practice is to make the radius of the link-arc equal to the length of the radius-rod; and when so made the lead is constant for all grades of the link. This property is at once evident from inspection of Fig. 1, Pl. XVIII, and Fig. 2, Pl. XVII, one having open and the other crossed rods; for it will be seen that when the engine is at a dead-point and the link is erect, the link-block may be moved from one end of the link-arc to the other without moving the valve.

Analytical Discussion.—Making use of the same notation as in the analytical discussion of the Stephenson link, let e be the eccentricity, and δ the angular advance for each eccentric; let c be the half-length of the link, and d the displacement of the link-block from the middle of the link; let l be the length of the eccentric-rod, the link-pins being assumed to be on the link-arc; let l_1 be the length of the radius-rod, and s the length of the valve-spindle.

Assume that the link is so suspended that m, the middle point of the chord, shall remain on the central line XX', and that the length of the link is sensibly the same whether measured on the chord or on the arc. Assume also that the rods are open.

In Fig. 2, Pl. XVIII, the diagram in fine lines represents the link-motion when the crank is on a dead-point; and the diagram in heavy line, represents it when the crank has moved from C_0 to C through the angle θ.

The distance from the origin O to the middle of the valve b is

$$Ob = Op - pk - kq + qS + Sb. \quad . \quad . \quad . \quad . \quad (37)$$

The term pk is determined by the equation

$$pk = KP \sin pPK = (c - a) \sin \alpha. \qquad (38)$$

Now

$$\sin \alpha = \frac{pp'}{PP'} = \frac{Op - Op'}{2c}. \quad \ldots \quad (39)$$

But

$$Op = Oe + ep = Oe + \{\overline{EP} - (Pp - Ee)^2\}^{\frac{1}{2}}; \quad (40)$$

$$Op' = -Oe' + e'p' = -Oe' + \{\overline{E'P'}^2 - (P'p' - E'e')^2\}^{\frac{1}{2}}. \quad (41)$$

in which

$$EP = E'P' = l, \qquad Pp = Pp' = c \cos \alpha;$$

$$Oe = r \sin (\theta + \delta), \qquad Oe' = r \sin (\theta - \delta);$$

$$Ee = r \cos (\theta + \delta), \qquad Ee' = r \cos (\theta - \delta).$$

Substituting these values in equations (40) and (41) gives

$$Op = r \sin (\theta + \delta) + \{l^2 - [c \cos \alpha - r \cos (\theta + \delta)]^2\}^{\frac{1}{2}}; (42)$$

$$Op' = -r \sin (\theta - \delta) + \{l^2 - [c \cos \alpha - r \cos (\theta - \delta)]^2\}^{\frac{1}{2}}. \quad (43)$$

A comparison of equations (42) and (43) with equations (12) and (16) shows that they differ in that the coefficient of $\cos \alpha$ does not contain d, and that only the upper sign of the double sign appears before the last term in the bracket; this last because the discussion applies only to open rods. Consequently the value of sin α may be obtained from equation (18) by omitting terms containing d, and using only the upper sign of the double signs; hence

$$\sin \alpha = \frac{r}{c}\cos \delta \sin \theta - \frac{r}{l}\sin \delta \sin \theta$$
$$- \frac{r^2}{4cl}[\cos^2(\theta+\delta) - \cos^2(\theta-\delta)]. \quad (44)$$

Expanding the term in equation (42) which has a fractional exponent, by the binomial theorem, and rejecting terms with higher powers of l in the denominator, and at the same time substituting unity for cos α, will give

$$Op = r\sin(\theta+\delta) + l - \frac{c^2}{2l} + \frac{cr}{l}\cos(\theta+\delta) - \frac{r^2\cos^2(\theta+\delta)}{2l}. \quad (45)$$

The first two terms of the equation (37) are now determined; the others are

$$Sb = s,$$

$$qS = \{l_1^2 - d^2\}^{\frac{1}{2}} = l_1 - \frac{d^2}{2l_1} \text{ (nearly)},$$

and

$$kq = \frac{c^2}{2l_1} - \frac{d^2}{2l_1} \text{ (nearly)}.$$

To obtain the last equation, it may be admitted that qk (Fig. 2, Pl. XVIII) is nearly equal to QK, which is nearly equal to tm. Now Pm is half of a chord bisected by a diameter, of which one segment is nm and the other is $2l_1 - mn$; consequently

$$Pm^2 = nm(2l_1 - mn);$$

$$\therefore\ mn = \frac{c^2}{2l_1} \text{ (nearly)};$$

and in like manner

$$nt = \frac{d^2}{2l_1} \text{ (nearly)}.$$

$$\therefore\ tm = kq = \frac{c^2}{2l_1} - \frac{d^2}{2l_1} \text{ (nearly)}.$$

Substituting the values of the several terms in equation (37),

$$Ob = r\sin(\theta + \delta) + l - \frac{c^2}{2l} + \frac{cr\cos(\theta + \delta)}{l} - \frac{r^2\cos^2(\theta + \delta)}{2l}$$

$$-(c-d)\left\{\frac{r}{c}\sin\theta\cos\delta - \frac{r}{l}\sin\theta\sin\delta\right.$$

$$\left. + \frac{r^2}{4cl}[\cos^2(\theta - \delta) - \cos^2(\theta + \delta)]\right\}$$

$$-\frac{c^2}{2l_1} + \frac{d^2}{2l_1} + l_1 - \frac{d^2}{2l_1} + s;$$

$$\therefore Ob = r\sin\theta\cos\delta + r\cos\theta\sin\delta + \frac{cr}{l}\cos\theta\cos\delta$$

$$-\frac{cr}{l}\sin\theta\sin\delta - r\sin\theta\cos\delta + \frac{cr}{l}\sin\theta\sin\delta$$

$$+\frac{dr}{c}\sin\theta\cos\delta - \frac{dr}{l}\sin\theta\sin\delta + l - \frac{c^2}{2l} - \frac{c^2}{2l_1} + l_1 + s$$

$$-\frac{r^2}{2l}\cos^2(\theta+\delta) - \frac{r^2}{4l}\cos^2(\theta-\delta) + \frac{r^2}{4l}\cos^2(\theta+\delta)$$

$$+\frac{r\,d}{4cl}\cos^2(\theta-\delta) - \frac{r^2d}{4cl}\cos^2(\theta+\delta);$$

$$Ob = r\left(\sin\delta + \frac{c}{l}\cos\delta\right)\cos\theta + \frac{dr}{c}\left(\cos\delta - \frac{c}{l}\sin\delta\right)\sin\theta$$

$$+ l + l_1 + s - \frac{c^2}{2l} - \frac{c^2}{2l_1}$$

$$-\frac{r^2}{4cl}[(c+d)\cos^2(\theta+\delta) + (c-d)\cos^2(\theta-\delta)]. \quad (46)$$

The last term is identical with the term dropped from equation (20), and it may be neglected here as well.

At the crank-end dead-point θ is zero, and it is 180° at the head-end dead-point. The corresponding values for Ob are

$$Ob' = r\left(\sin\delta + \frac{c}{l}\cos\delta\right) + l + l_1 + s - \frac{c^2}{2l} - \frac{c^2}{2l_1}; \quad (47)$$

$$Ob'' = -r\left(\sin \delta + \frac{c}{l}\cos \delta\right) + l + l_1 + s - \frac{c^2}{2l} - \frac{c^2}{2l_1}. \quad (48)$$

Hence the distance from the origin O to the middle of the valve, when in mid-position, is (Fig. 2, Pl. XVIII)

$$Oo = \tfrac{1}{2}(Ob_1 + Ob_2) = l + l_1 + s - \frac{c^2}{2l} - \frac{c^2}{2l_1}. \quad (49)$$

The displacement of the valve from mid-position at any crank-angle is obtained by subtracting equation (49) from equation (46), member from member, giving

$$e = r\left(\sin \delta + \frac{c}{l}\cos \delta\right)\cos \theta + \frac{dr}{c}\left(\cos \delta - \frac{c}{l}\sin \delta\right)\sin \theta. \quad (50)$$

Thus far in this discussion attention has been given to the case of open rods only; were the same method to be carried through for crossed rods, using a figure similar to Fig. 2, Pl. XVIII, a similar equation would be found for the valve-displacement, except that the quantities c and d would be affected by a negative sign.

Zeuner's Diagram.—A comparison of equation (50) with equation (29) shows that a valve controlled by a Gooch link-motion has a harmonic motion, and that its displacement from mid-position, at any grade of the link, may be represented by a valve-circle. Taking account of the observation at the end of the previous paragraph, with regard to crossed rods, the co-ordinates of the end of the diameter of the valve-circle may be written

$$A = r\left(\sin \delta \pm \frac{c}{l}\cos \delta\right); \quad \ldots \quad (51)$$

$$B = \frac{dr}{c}\left(\cos \delta \mp \frac{c}{l} \sin \delta\right); \qquad (52)$$

the upper sign being used with open and the lower with crossed rods.

The expression for A, the abscissa of the end of the diameter of the valve-circle, is the same for all grades of the link; which agrees with the statement on page 54, that the lead is constant when the radius of the link-arc is equal to the length of the radius-rod. It is apparent, therefore, that a Gooch link-motion is equivalent to a shifting eccentric with constant lead. Fig. 3, Pl. XVII, gives the valve-circles for full-gear, mid-gear, and for two intermediate gears; and shows the variation of cut-off from full-gear to mid-gear. As was found to be the case with the shifting eccentric, constant lead is found to be accompanied by an earlier admission from full-gear toward mid-gear, though the change is not so marked as it is with an increasing lead.

It must be noted that the Gooch link-motion at full-gear does not give the valve the motion that it would have if the connection were made by an eccentric-rod to the head of the valve-spindle. In this respect the action of the Gooch link-motion differs from the Stephenson link-motion, which at full-gear acts like a plain slide-valve gear. The diameter of the valve-circle is, at full-gear,

$$(A^2 + B^2)^{\frac{1}{2}} = r\left(\sin^2 \delta + \frac{c^2}{l^2}\cos^2 \delta \pm 2\frac{c}{l}\sin \delta \cos \delta\right.$$

$$\left. + \cos^2 \delta + \frac{c^2}{l^2}\sin^2 \delta \mp 2\frac{c}{l}\sin \delta \cos \delta\right)^{\frac{1}{2}}$$

$$= r\left(1 + \frac{c^2}{l^2}\right)^{\frac{1}{2}},$$

which for the ordinary proportions of the link-motion is a trifle longer than r. Consequently the full-gear action of a Gooch link-motion with open rods is equivalent to that of a plain slide-valve gear with a little greater angular advance; with crossed rods it is equivalent to the action of such a gear with a little less angular advance. The difference in each case, though not large, is appreciable.

The Allan Link.—At the time when link-motions were first used, the curved surfaces of either the Stephenson or the Gooch link could be properly finished only with consiaerable difficulty and expense. To obviate this difficulty, a straight link was devised by Allan which had the general appearance of the Gooch link, but which had both the link and the radius-rod movable in such a way as to give a proper motion to the valve. This gear was intermediate between the Stephenson and the Gooch link-motion; for example, it had a variable lead, though the variation was less than with a Stephenson link having like proportions. With modern machine-tools and shop methods, there is no especial difficulty in finishing the curved surfaces of links, and the Allan link has consequently fallen into disuse.

Comparison of the Stephenson and Gooch Links.— A comparison of the link-motions on Plates XIII and XIV with that on Plate XVIII will show that the Gooch link-motion has more parts and more joints at which lost motion will result from wear, and that it occupies nearly twice the longitu- dinal space required for a Stephenson link-motion. As an offset may be urged its property of giving a constant lead. The choice of a link-motion for a specific purpose must depend on the importance that should be attached to any given feature of the gear under the given conditions. With proper proportions either gear can be made to give the valve a nearly harmonic motion, or, with proper modifications, either gear may be adjusted to give an equalized cut-off; from this point of view neither appears to have an advantage.

A link-motion is used primarily to reverse the motion of the engine, and secondarily to give a variable cut-off; under some conditions the second action is nearly if not quite as important as the first. At one time link-motions under the control of a governor were used to give an automatic cut-off on stationary, non-reversing engines; and for such a purpose the Gooch link-motion was preferable because it imposed less friction on the governor. The weight of the Stephenson link-motion can be overcome either by counterbalancing or by speeding up the governor, and mass is rather advantageous than otherwise in that it steadies the governor; but the friction produced by the weight of the link and the eccentric-rods, and much more the friction of the gear, which is liable to be large and irregular, especially at the eccentrics and their straps, ought not to be imposed on a governor. The constant lead of the Gooch link-motion has been claimed as an advantage, and engine-builders usually so consider it; but the discussion of this question in connection with shifting eccentrics, on page 37, makes it clear that a decreasing lead is preferable for a stationary engine which has a variable cut-off with a single valve. Were this the only point under consideration, the best result could be obtained by a Stephenson link-motion with long crossed rods, which would give a lead decreasing slowly toward mid-gear. Modern builders of stationary engines do not look favorably on the link-motion for an automatic cut-off gear.

Reversing engines are of two types, those used on locomotive engines and those used on marine engines; the conditions of their service are so different as to merit detailed discussion. Some stationary engines are reversing engines, and are controlled by hand instead of by a governor; for examples may be mentioned winding and hoisting engines, and engines for driving reversing roll-trains. According to the conditions of their use, they will fall into one or other of the two classes mentioned, or may partake of the characteristics of both.

In starting a railway train, the link-motion is thrown into full-gear forwards, and should then give a long cut-off, so that, with the throttle-valve partially open, a moderate and steady force may be exerted on the driving-wheels to overcome the friction of, and impart motion to, the train, without slipping the wheels on the track. The lead may properly be small at full-gear, and is sometimes zero or even negative. As the train gets under way and the revolutions per minute become high, the action of the reciprocating parts becomes important; just as was seen to be the case for a high-speed stationary engine (see page 36) there must be a considerable amount of compression in order that the engine may run smoothly. An early admission and release are also desirable in order that the steam may be supplied and exhausted freely. All these conditions are met by raising the Stephenson link toward midgear and opening the throttle-valve, and at the same time the economic advantage of the expansive working of steam can be obtained.

American and English locomotive-designers have used the Stephenson link-motion, while Continental designers have used the Gooch link quite widely. In American practice the cylinders are commonly placed outside of the locomotive-frames, with the valve-chest on top; the link-motions are placed between the frames and act on the valve through a rocker, as has already been shown on Plate XIII. English locomotives frequently have the cylinders inside the frames, and the valve-chests are on the sides of the cylinders and under the smoke-box; the link-motions then act directly on the valve-spindle, using a link like that shown by Fig. 3, Plate XIII. When the Continental locomotive-designers use the Gooch link-motion, they place it outside the drivers, and so readily find room for eccentric-rods and radius-rod. Such a disposal of the valve-gear keeps it in sight where it may receive attention, but does not meet with favor among American and English engineers, since it is liable to derangement from slight accidents.

At the present time marine engines are commonly compound or triple-expansion, and the link-motions are used only for reversing the engine, the reverse arc having only three notches, full-gear forward, full-gear backing, and mid-gear. Crossed rods may be used to advantage, for then the engine can be stopped by setting the link at mid-gear. In this connection it may be remarked that an engine controlled by a Stephenson link-motion with open rods will not necessarily stop when the link is placed in mid-gear, provided the engine is running under no load or a very light load; though the engine will not start with the link in that position. The normal condition for a marine engine is to run at full speed and under a full load, and when the speed decreases the load falls off rapidly. The result of an attempt to adjust the steam-supply to a smaller load, by shifting the link toward mid-gear, is to give an excessive compression, and an early release, when the reciprocating parts have least effect, and when the reduced quantity of steam is readily exhausted. Simple-expansion engines and many compound engines were provided with an independent cut-off valve on the back of the main valve, of a type to be discussed in Chapter V.

Locomotive link-motions have the pins and other smaller wearing parts made of hard steel, the link is case-hardened, and the eccentric-rods are bushed with steel; the eccentrics and straps are the only exception to the rule that all the wearing parts are made as hard as possible. With the exception of the eccentric-straps, no provision is made for taking up wear; when the looseness becomes excessive the whole gear is overhauled, and new pins and bushings provided if necessary. The reason for this practice is twofold: first, the complication of adjustable parts is avoided; second, the gear is unavoidably exposed to dirt and grit, and when grit gets into a joint between a hard and a soft metal, it becomes embedded in the latter and rapidly abrades the hard surface. The custom is to equalize the cut-off by a method that gives a good deal of slipping of the block in the link, but as both link and block are hard, and the cut-off is continually varying, this practice is not so objectionable as it would be on a marine-engine link-motion.

Marine-engine link-motions are designed to give the required cut-off for full load when set at full-gear; as has already been said, there is frequently no provision for shortening the cut-off as on locomotives. Since the engine is liable to run for days or weeks at full-gear forward, the gear is designed to give very little slipping of the block in the link at this position. As the gear is large and frequently massive, the complication of making the wearing parts adjustable is not objectionable; and as the engine works in a closed engine-room, there is no reason why grit should get into the wearing surfaces, which may therefore be lined with soft metal when that is desirable.

Modification of the Link-motion.—In the discussion of link-motions, hitherto, it has been supposed that the arrangement of the parts and the choice of dimensions have been such as to give a nearly harmonic motion to the valve. In both the Stephenson and the Gooch link-motions the link-pins are supposed to be on the link-arc; in the first the radius of the link-arc is assumed to be equal to the length of the eccentric-rod, and in the second it is assumed to be equal to the length of the radius-rod. It is, however, possible by modifying some of the arrangements to obtain certain desired effects, such as the equalization of the cut-off, without sacrificing the equality of the leads. The effect of some of the modifications can be proved, or at least inferred, from the diagram of the link-motion; but since they are in the nature of adjustments and must in any case be worked out by trial, it will be sufficient to state some of them for the Stephenson link-motion.

Link-pins.—The link-pins of a Stephenson link-motion *with a rocker* may advantageously be placed back of the link-arc; *without a rocker* they should be on the link-arc, or they may be placed ahead of the arc if mechanical difficulties do not interfere.

Saddle-pin.—By proper location of the saddle-pin, or the point of attachment of the hanger or bridle, it is possible to equalize the cut-off at any point of the stroke both in forward and backing gears. This element of the link-motion is by far the most effective, for good or evil, of all that the designer has at his control, and fortunately he usually has complete control over it. If a symetrical gear is desired, the saddle-pin should be placed at the middle of the length of the link; *with* a rocker it will be back of the link-arc, and *without* a rocker it will be ahead of the link-arc. With proportions common for American locomotives a fair action may be had by placing the saddle-pin at the middle of the link and halfway between the chord and the arc. On such locomotives, with a rocker, the slip of the link-block is greater in forward than in backing gear, when the saddle-pin is at the middle of the link; and the forward-gear slip may be diminished by placing it nearer the forward-gear eccentric, but this is attained at the expense of the symmetry of the gear. Sometimes the link is supported from below instead of being suspended from above; and in such case the forward-gear slip is less than the slip in backing gear. Links for marine engines, and in English locomotives without a rocker, are frequently suspended by the forward link-pins. It is customary to equalize the cut-off at half-stroke or earlier, by a proper location of the saddle-pin; in this book an equalization at one-third stroke will be made.

Reverse-shaft.—The position of the reverse-shaft is often fixed, or susceptible of but little change. If it can be located at pleasure, it may also be used to equalize the cut-off at any point of the stroke. When so used, the location of the reverse-shaft is used to equalize the cut-off near the end of the stroke, usually in combination with an equalization of the cut-off by the location of the saddle-pin at half-stroke or earlier. With a rocker, such a manner of locating the reverse-shaft is liable to bring it in conflict with the eccentric-rods at one of the full-gear positions.

Radius of the Link-arc.—It has been shown that the link-arc for the Stephenson link should have a radius equal to the length of the eccentric-rod, in order that the leads may be equal. It may sometimes be desirable to use a different radius to facilitate the equalization of the cut-off. *Without* a rocker, the radius may be made greater than the length of the eccentric-rod, and *with* a rocker it may be made less. Such a choice of the radius of the link-arc will sacrifice the equality of the leads, and the deviation from the normal radius must never be enough to badly derange them. In this connection it should be said that the lead near mid-gear supplies a large portion of the steam admitted, and that the lead at full-gear affects the facility with which the engine passes the centres; much inequality in either place is undesirable. With open rods, the full-gear lead is small, and may vary from a certain amount to double that amount, or to zero, without serious consequence. At and near mid-gear the lead is large, and may vary as much absolutely as at full-gear, since that will not be much relatively.

The motion of the valve will be more nearly harmonic with long rods and with a long link; the first should be twelve times and the second four times the eccentricity, or more, except under peculiar conditions. A skilful designer may use the inequality introduced by short rods or a short link to adjust a link-motion, but these dimensions are commonly fixed, and the modification of them for that purpose is not in general to be regarded with favor.

Designing Link-motions.—The design of a link-motion may be divided into two parts: first, the choice of such a type of link-motion and such general proportions as will be likely to give a satisfactory solution of the problem in hand; and, second, the application of modifications and adjustments to give equal cut-off, or to reduce the slip, or to produce any other desired effect. The first part of the design may be much aided by the use of the Zeuner diagram; the second part is commonly attained either by drawing out the link-motion and by making the proper constructions, or by aid of a working model with adjustable parts. A combination of the methods of drawing and using a model, combining certain advantages of each, will be explained later. The first part of the design is often so fixed by the requirements of the general design of the engine, or by custom, that it is liable to be neglected.

Marine-engine Link-motions.—As has already been stated, the link-motions of modern marine engines are commonly used for reversing only, and not for regulating the power of the engine. Since the Stephenson link-motion at full-gear acts like a plain slide-valve gear, the determination of the eccentricity and angular advance and the design of the valve may be carried out by the methods given in the first chapter. It is customary to make the head-end lap greater than the crank-end lap, thereby partially equalizing the cut-off, and the consequent inequality of leads (the crank-end lead being the larger) is a partial compensation for the longer cut-off at the head end. In Fig. 1, Pl. V, the only change required is to choose the point of cut-off for both forward and return strokes, the former being the longer.

The link is usually guided by the go-ahead pin, as shown on Plate XIV and Plate XIX, both of which represent the link-motion of the U. S. S. *Maine*; though sometimes a point beyond the link-pin, or a point at the middle of the link, is chosen. This particular link-motion actuates a piston-valve like that shown by Fig. 1, Pl. VII, which takes steam in the middle and exhausts at the ends. Consequently the motion of the valve is in the opposite direction from that of a plain slide-valve, and the general effect is like that produced by moving a valve through a rocker; the eccentric follows the crank, and the rods are crossed, as shown on Plate XIV and Plate XIX. Let p and P be the positions of the go-ahead link-pin when the crank is on crank-end and head-end dead-points respectively; then with p and P as centres and with a radius equal to the length of the bridle, arcs are struck intersecting at N, which is one extreme pósition of the end of the reverse-shaft arm. The length of the bridle is such that the arc $aPpa_1$ nearly coincides with the line XX', and the slipping of the link-block is small; the arc aa_1 is extended both ways for sake of clearness in the diagram.

The location of the reverse-shaft may now be made either to give the backing action symmetrical with the forward action of the link, or to reduce the slip in backing-gear.

Symmetrical Action.—Let it be assumed that the backing-link pin shall be guided to the points P and p, when the engine is on the dead-points; then the go-ahead link-pin will be found at P_1 and p_1. With these points as centres and with the length of the bridle for a radius, draw arcs intersecting at n; and with N and n as centres and with the length of the reverse-shaft arm as a radius, draw arcs intersecting at S: the last point is the desired location of the reverse-shaft.

Reduction of the Slip.—In full-gear forward the go-ahead link-pin moves on the arc aa_1, and the backing link-pin describes an elongated looped figure, of which the upper loop is quite small; P' and p' are two points on the larger loop. If now the link be thrown into full-gear backing, and if the backing link-pin be made to move on the line XX', then the go-ahead link-pin will describe a looped figure of which P_1 and p_1 are two points. To draw the looped figure, let C be a given position of the crank, and let E and E' be the corresponding positions of the eccentric-centres; with E' as a centre and with the length of the eccentric-rod for a radius, cut the line XX' at P; with E as a centre and with the same radius, draw an arc and intersect it with another arc drawn from P and with the length of the link between the pins for a radius; then P_1 at the intersection of the two last arcs is one point on the looped figure. To find other points, make the same construction for a sufficient number of crank-positions, say twelve, at equal intervals around the circle described by the crank-pin.

Find by trial a centre n_0, from which, with a radius equal to the length of the bridle, the arc tt_1 may be drawn through the middle of the looped figure. Then with a radius equal to the length of the reverse-arm, draw arcs intersecting at S_0; this is the location of the reverse-shaft for giving as small a slip as possible in full-gear backing. With this construction the go-ahead link-pin describes the arc tt, when the link is in full-gear, and at the same time the backing link-pin describes a looped figure similar to the one at $P'p'$, and lying partially on one side and partially on the other side of the line XX' near Pp. The link-block will slip in the link an amount nearly equal to the width of the looped figure P_1p_1, measured on a radial line from n_0; the exact amount of slip can be found by drawing the true looped figure near Pp; it is omitted in the diagram to avoid confusion. It is evident that the slip of the link-block will be greater if the construction for symmetrical action resulting in the location of the reverse-shaft at S, should be used. In that case the slip is nearly equal to the total deviation of the looped figure P_1p_1 from the arc through P_1p_1; it can be found by drawing the true looped figure near Pp when the path of the go-ahead link-pin is the arc P_1p_1.

Adjusting the Cut-off.—The distribution of work among the cylinders of a compound or multiple-expansion engine depends on the ratio of the volumes of the cylinders and on the cut-off for the several cylinders. If the distribution of the work is not satisfactory when all the links are set at full-gear, it may be adjusted, or at any rate it may be improved by shortening the cut-off on one or more of the cylinders. On recent marine engines, which have only three notches on the reverse-arc, namely, full-gear forward, full-gear backing, and mid-gear, a device known as a gag is put on the end of the reverse-arm for this purpose. The rod of the bridle is carried by a block N (Fig. 1, Pl. XIV), which may be moved in a slot

NM in the end of the reverse-shaft arm, by aid of a screw and hand-wheel, and thus the link may be moved toward mid-gear and the cut-off may be shortened.

The construction of the gag for the link-motion shown on Pl. XIV is found on Pl. XIX. A line is drawn midway between nP_1 and np_1, and perpendicular thereto is drawn the line nm; this is chosen as the centre-line of the slot and screw NM, Fig. 1, Pl. XIV. In full-gear forward the centre-line of the slot and screw is at MN, which makes an angle of about 7° with a line midway between NP and Np. It is apparent that when the block N is moved toward M, Pl. XIV, the cut-off is shortened for forward-gear without changing the method of supporting the link-motion, while in backing gear the cut-off is not affected. If the slot should be placed at $M'N$, $m'n$, making the angle PNM' equal to $180° - P_1nm'$, then the cut-off will be shortened the same amount in both forward and backing gears; but with such a construction the method of supporting the link in forward gear will be affected by the action of the gag, and slipping of the link-block may then occur.

Locomotive Link-motions.—In American practice the link-motion is set to give equal lead at full-gear, and is adjusted to give equal cut-off; the reduction of slip being considered to be of less importance, though it is not to be neglected. The adjustment of the cut-off is made by aid of a model with adjustable parts, by aid of which an experienced designer can readily work out a satisfactory motion, or at any rate as good a motion as the conditions will allow.

In equalizing the cut-off, it is to be borne in mind that any irregularity at short cut-off is of much more importance than at long cut-off, since the amount of steam admitted to the cylinder is nearly proportional to the length of the cut-off, and the work varies with the amount of steam admitted. Thus an inequality of half an inch in six inches is $\frac{1}{12}$, while half an inch in eighteen inches is only $\frac{1}{36}$.

The full-gear or maximum cut-off varies with the condi-

72 VALVE-GEARS FOR STEAM-ENGINES.

tions of the service and the judgment of the designer from $\frac{3}{4}$ to $\frac{11}{12}$ of the stroke. When an engine, for example one in passenger service, is to be run at high speed and with a short cut-off, it is advisable to make the full-gear cut-off as short as the ready handling of the engine, when starting, will permit, in order that a favorable action of the valve may be obtained at short cut-off. This will be made clear by a comparison of Fig. 4, Pl. XVI, in which the maximum cut-off is at $\frac{2}{3}$ of the stroke, with Fig. 1, Pl. XX, in which the maximum cut-off is at $\frac{5}{8}$ of the stroke; the diameters of the full-gear valve-circles are the same. In each figure OP_1 is the valve-circle for the link-block half-way between the link-pin and the saddle-pin. In Fig. 4, Pl. XVI, OP_1 has a diameter of $\frac{29}{32}$ of an inch, gives a maximum port-opening of $\frac{15}{32}$ of an inch, and the cut-off occurs at 0.70 of the stroke. In Fig. 1, Pl. XX, the cut-off occurs at 0.55 of the stroke, the diameter of OP_1 is $1\frac{1}{32}$ of an inch, and the maximum port-opening is again $\frac{15}{32}$ of an inch. The comparison also shows the importance of the first part of the design of a link-motion, mentioned on page 67, and the advantage of using Zeuner's diagram for that purpose.

Skeleton Model.—The author has found that a skeleton model, such as is shown on Pl. XX, can be used with advantage in laying out and adjusting a link-motion. It consists of a piece shown by Fig. 3 to represent the crank and eccentrics, of a template shown by Fig. 4 to represent the link, and of rods to represent the eccentric-rods, the hanger and the reverse-shaft arm, together with screws and washers to make attachments. The several parts may be made of any fine-grained hard wood, such as mahogany or cherry. Fig. 5 shows one of the joints with a thick wooden washer, as at the eccentric-centre e'. A wood screw of proper size is cut off so that it may not protrude through the plate into which it is set. The hole at b for the screw is drilled a trifle smaller than the shank of the screw at the bottom of the threads, and a pointed screw like the one to be used is run through to cut a thread in the hole. The hole in the washer c may be an easy fit for the screw, while the hole in the rod d is a snug fit for the body of the screw under the head. The countersunk hole in the rod d should be made with a tool like a machinist's countersink, but with the cutting edges sharper. The drilling and countersinking should be done in a lathe, or by some other method that will give true and straight holes. After the holes are laid out, a sharp-pointed centre-punch may be used to start the drill at the proper point. All the work on the model should be done exactly and thoroughly.

In anticipation of laying out the model, a Zeuner's diagram should be drawn as in Fig. 1; in which the full-gear valve-circle OP is laid out as for a plain slide-valve to give the lead-angle XOR_a and the cut-off at OR_1 corresponding to the piston-position a, on the assumption of harmonic motion. As was found to be the case for a plain slide-valve, it may be necessary to modify and redraw the full-gear valve-circle in order to get a desired lead and lap. The diameter of the mid-gear valve-circle may be conveniently calculated by aid of the equation (33), and then the locus $PP_1P_2P_0$ of the end of the valve-circle diameters may be drawn as the arc of a circle centred on the line XX'.

The crank template (Fig. 3) may now be laid out by drawing the axes OR and YOY, and laying out E and E' with the eccentricity and angular advance found in Fig. 1. At R a slip of brass is let into the under side of the template, or a slip of card-board is gummed onto the under side; and the axis OR is drawn across the slip for a reference-mark. If card is used, it is well to gum a piece to the centre at O, so that the template may turn smoothly. The line OR represents the crank, and may be conveniently six inches long for all models whatever the true length of the crank of the engine. Holes are drilled at E and E' to receive screws, and a hole at O is drilled and countersunk for a screw-head.

The link template (Fig. 4) has the arc jk cut to the radius of the link-arc. *mq* is a diameter at the middle of the arc, and *jp* and *kp'* are parallel lines on which the link-pins *p* and *p'* are laid out, at a proper distance back of the arc. Holes are drilled at *p* and *p'* to receive screws.

The rods which serve as eccentric-rods, hanger, and reverse-shaft arm are laid out with the proper lengths, and are drilled and countersunk as may be required. The latter is made wide enough to receive a weight to hold it in place during the use of the model.

Washers will be required at *e'*, *p'*, *m'*, and *N*, Fig. 2.

On a drawing-board or drawing-table of proper size, spread a sheet of paper large enough to receive the model, and make constructions shown by Fig. 2. If desired, two smaller sheets may be used, one for the crank and eccentrics, and one for the link and the path of the cross-head. On the paper lay down a line *XX'* for the axis of the drawing, preferably by aid of a long steel straight-edge. From the centre *O* draw a circle to represent the path of the crank, with a radius equal to *OR*, Fig. 3. With a pair of beam compasses take off the length of the connecting-rod to the same reduced scale as is used for the crank *OR*, and with the ends of the diameter R_0 and R_0' for centres cut the axis *XX'* at *x* and *x'*, the ends of the path of the cross-head pin. Divide the path of the cross-head into fractional or decimal parts, or into inches, to the chosen reduced scale, as may be desired. Connect the eccentric-rods to the crank template and the link template, reserving the hanger and reverse-shaft arm till the saddle-pin is located.

Full-gear Lap and Lead.—Set the model at a dead-point with the link-pin *p* on the axis *XX'* as shown by Plate XIII, and mark the intersection of the link-arc with the axis *XX'*; place the model on the other dead-point, *p* on *XX'*, and again mark the intersection of the arc with the axis; thus locating the points *c* and *c'*. Since the valve-displacement at a dead-point is equal to the lap plus the lead, *cc'* is twice *Oc*, Fig. 1. Make *on* equal to *on'* equal to the lap; then, whenever the link-arc is on one of the points n or n', the displacement of the valve is equal to the lap, and it is either at admission or at cut-off.

Location of the Rocker.—The axis of the rocker is on a line oT perpendicular to the axis XX' at the point o. In order that the admission and cut-off may not be disturbed by the action of the rocker, the axis T may be so chosen that an arc from T shall pass through n and n'. For convenience in laying out and erecting the engine, oT may be made equal to the length of the arm of the rocker; in that case n and n' are on an arc drawn through o, and at a horizontal distance from oT equal to the lap.

Location of the Saddle-pin.—The location of the saddle-pin is commonly under the entire control of the designer, and should be so chosen as to give the best general action of the link, bearing in mind that an inequality at short cut-off is more deleterious than at long cut-off. It will in general be found advisable to equalize the cut-off at or near one-third stroke by aid of the saddle-pin.

In order that the cut-off may be equalized at one-third stroke by this means, a preliminary location of the reverse-shaft is necessary; this may be made by placing S, the reverse-shaft axis, at a distance above XX' equal to the length of the hanger, and at a distance to the left of o equal to the length of the reverse-shaft arm. On Plate XX S is a little further to the left for a reason to be explained later. From the preliminary location of the reverse-shaft as a centre draw an arc, such as N_1N_2.

With the length of the connecting-rod, to the reduced scale, as a radius, and with points at one-third stroke on xx' for centres, draw arcs cutting the crank-pin circle at R and R', to find the crank-positions corresponding. Bring the reference-mark on the crank template to one of these points, as at R' for the return stroke, and put a weight on the template to hold it in place. Slip the link template up or down on the paper till the link-arc comes to the point n', and mark the extremities of the line $m'q'$ on the paper and draw a bit of the arc; drop the link template out of the way and draw a line connecting the points m' and q'. This part of the construction is indicated on Plate XX, with the rods and templates in full lines. Bring the reference-mark of the crank template to R, and slide the link template up or down till the link-arc comes to n; mark and draw the line mq together with a bit of the arc. This part of the construction is drawn on the plate with dotted lines, showing the link template and the centre-lines of the eccentric-rods.

With the length of the hanger as a radius, find by trial a point on the arc N,N_1 from which an arc may be drawn which shall intersect the lines mq and $m'q'$ at points m and m', at equal distances from the portions of the link-arc drawn at the forward extremities of those lines. This part of the work must be done with great nicety, and the surface of the paper should not be defaced by drawing unnecessary lines.

Having the distance of the saddle-pin back of the link-arc, the location may be laid out on the link template, and a hole drilled to receive the screw at the end of the hanger.

Location of the Reverse-shaft.—Though the location of the reverse-shaft, as has already been stated, may be used as an element for the adjustment of the link-motion, it is frequently out of the power of the designer to change it enough to produce much effect, and in general a good action of the link may be had without doing so. A small change in the location of the reverse-shaft has little effect on the action of the link; consequently it may be made to suit the necessities or convenience of the general design of the engine, within limits; for example, it is convenient to have S at some definite distance from O, measured in inches and such fractions of an inch as are in common use in the shop.

On Plate XX the reverse-shaft is so located that the hanger may swing through equal angles forward and back of the vertical, when the link is in mid-gear. This is accomplished by making oi equal to the distance back of the link-arc, and then by drawing in_o and n_oS vertical and horizontal, and equal to the lengths of the hanger and reverse-shaft arm respectively. When the reverse-shaft is thus located, the path of the saddle-pin at full-gear is not level, but is inclined upward a little at the end nearest the crank, something like the arc $m_a'm_a$ Fig. 2, Pl. XXI, though to a less degree. This is beneficial so far as the cut-off is concerned, though it is likely to increase the slip. It is probable that the cut-off may be further improved in some cases by setting the reverse-shaft still further towards the axle O. Again, the reverse-shaft may be set further forward and so distribute the inequality due to the curvature of the path n_1n_2 of the upper end of the hanger.

Finally, it may be stated that in general it will be sufficient to locate the reverse-shaft as far above the axis XX' as the length of the hanger, and as far back of o, the mid-position of the link-block, as the length of the reverse-shaft arm; or the preliminary location already stated may be made definite at the beginning.

If it is considered advisable to use the location of the reverse-shaft to adjust the link-motion, then it may be made to equalize the cut-off at some point near the end of the stroke (for example, at ⅔ of the stroke) after the location of the saddle-pin has been used to equalize the cut-off at or near ½ of the stroke. The construction is as follows: Place the reference-mark on the crank template at R_a, Fig. 2, Pl. XXI, the crank-position corresponding to ⅔ stroke forward, and bring the arc of the link-template to the point n; then mark the line m_aq_a and the position m_a of the saddle-pin. Place the reference-mark of the crank template at R_b, the crank-position corresponding to ⅔ return stroke, and bring the link-arc to the point n'; then mark the line $m_a'q_a'$ and the location of the saddle-pin m_a'. Make similar constructions in backing gear, thus locating the points m_b and m_b', the positions of the saddle-pin at cut-off at ⅔ stroke for the forward and return stroke; or, since the diagram in backing gear will be symmetrical with that for forward gear, the points m_b and m_b' may at once be laid off above the axis XX' symmetrical with the points m_a and m_a'. With the length of the hanger for a radius, and with m_a and m_a' as centres, draw arcs intersecting at N_a; and with the same radius and with m_b and m_b' as centres, draw arcs intersecting at N_b. Again, with the length of the reverse-shaft arm as a radius and with the points N_a and N_b as centres, draw arcs intersecting at S'; this is the desired location of the reverse-shaft. It is to be noted, first, that such a location of the reverse-shaft is liable to derange the cut-off at $\frac{1}{4}$ stroke, and that consequently the saddle-pin must be relocated; and second, that the axis of the reverse-shaft is thrown down so that it is liable to conflict with the eccentric-rod.

Lead, Port-opening, and Slip.—After the locations of the saddle-pin and reverse-shaft are completed, the model is then to be used to test the action of the link in all grades forward and backing. To do so, connect up the reverse-shaft arm and the hanger; place the reference-mark on the crank-template, at the crank-position corresponding to a chosen piston-position (for example, at half-stroke forward); adjust the link-template so that its arc shall come to the point n, Fig. 2, Pl. XX; and place a weight on the reverse-shaft arm. Turn the crank template and thus move the model till the link template comes to the point n'; note the position of the reference-mark on the crank template, and find and record the corresponding position of the cross-head on its path xx'.

With the same setting of the model find the lead, port-opening, and slip. The first will be found by placing the crank on the dead-points. The second must be found by noting the greatest displacement of the link-arc from n toward the right, and from n' toward the left; these quantities are to be measured on a horizontal chord, not on the arc, since they represent the valve-displacements minus the lap. To find the slip, find and mark the highest point on the link-arc that comes to the arc through nn', and also the lowest point; the distance between these points is the slip of the link-block.

This investigation of the action of the link should be carried on systematically for a sufficient number of points of the stroke, and recorded in a table similar to that on page 82.

Should the action of the link-motion be deemed unsatisfactory, experiments may be made by changing the various dimensions of the link, so far as possible, following the general directions on page 65. It is probable that the most troublesome element will be the slip of the link-block due to the attempt to equalize the cut-off, and with the method of suspension shown on Plates XIII and XX this is worse in forward than in backing gear. When feasible, this unfortunate condition may be reversed by supporting the link from below, with which arrangement the slip in forward gear will be less than in backing gear.

Modifications of the Model.—The model as described is made in an inexpensive and yet a thoroughly serviceable manner, and with proper care may be kept in constant use for several days before the joints begin to show looseness. If conditions warrant, with a little care in design and expense in construction, and with the use of metal bearings and of devices for taking up wear, a model can be made, on the same general plan, that may serve for an indefinite time. It does not appear necessary to go into the details of such changes at this place.

One of the most obvious changes, and one that is necessary if the pin is to be re-located, will be to provide some way of adjusting the saddle-pin; for example, it may be placed on a block that can be slid and clamped on the link template.

In certain positions of the model the eccentric-rod ep, Fig. 2, Pl. XX, strikes the washer at e'. This difficulty may be partially removed by cutting a notch in the rod as at v, and by cutting the washer flat on one side as shown by the line ab, Fig. 5. If desired, another notch may be cut at w, and the rod may be reversed end for end, thereby still further removing this source of trouble; but the screws at e and p are liable to become loose if they are taken out and reset many times.

In case the link-pins are on the link-arc, the attachment of the rods may be made by aid of brass plates, screwed to the link template and drilled for screws that must be set into the ends of the eccentric-rods.

If the link is supported from the go-ahead link-pin, as is common in English locomotive practice, then the conditions will be found to be similar to those for a marine-engine link-motion shown on Plate XIV, and the equalization of the cut-off, if attempted, must be made by the location of the reverse-shaft.

A skeleton model may readily be devised for adjusting and investigating the action of the Gooch link-motion.

To Set a Link-motion.—If a link-motion is designed for equal lead, it is set by a method like the first method for setting a slide-valve with equal lead. Place the link at full-gear forward; with the proper angular advance as near as may be for each eccentric, and with the engine on a dead-point give the valve the proper lead; turn the engine into the other dead-point, and if the valve does not then give the proper lead, change the length of the eccentric-rod by half the error, and shift the eccentric till the proper lead is obtained. Place the link at full-gear backing and set the valve again, changing the length of the eccentric-rod and the angular advance as may be necessary. Now place the link again at full-gear forward, and see if the setting has been disturbed by the changes of the backing eccentric and eccentric-rod; if it has, the valve must be reset by the same method. Place the link again in full-gear backing, and make any correction needed; very commonly none will be required. When a link-motion is designed to give unequal leads (a common practice on marine engines), the process differs only in that the lead at each end must be made the proper amount.

The second method given on page 30 cannot be used in

setting a link-motion, since the maximum port-openings are not likely to be equal.

When a rocker is employed with a link-motion, the length of the valve-spindle should be such that the arm of the rocker shall swing equal angles on each side of a perpendicular through its axis, to the central line or axis of the link-motion; but a little divergence from this condition will not have much effect.

The eccentric-rods for the link-motions on Plates XIII and XIV are joined to the eccentric-straps by T-heads and bolts. The length of the rods may be adjusted by placing slips of metal called shims under the head—a method which may appear crude, but is really convenient and effective. The bolt-holes through the shims can be slotted through to one side in order that a shim may be put in or withdrawn without taking the bolts out.

Application of Skeleton Model.—To illustrate the method and to show the influence of changing some of the parts of a link-motion, there are given here the results of the application of the skeleton model to a few examples.

The following dimensions were used in all the examples:

Eccentricity	3	inches.
Outside lap	$\frac{3}{4}$	"
Radius of link-arc	48	"
Length of hanger	16	"
Length of reverse-arm	20	"
Length of rocker-arm (when used)	12	"
Stroke	12	"
Ratio of crank to connecting-rod	1 : 5	

In all the examples except that shown by Table II the distance from the centre of the eccentric to the link-arc was equal to the radius of the link-arc; the length of the eccentric-rods from the centre of the eccentric to the link-pins was made less or more than this distance according as the link-pins were back of or ahead of the link-arc, except for the example shown by Table V, which had the link-pins on the link-arc. The saddle-pin in all cases is at the middle of the length of the link. Other dimensions of the link-motion and details of the arrangement are given with the tables.

TABLE I.

Rocker used—link-pins $2\frac{7}{8}$ inches back of link-arc.
Cut-off equalized at $\frac{1}{8}$ stroke.
Saddle-pin $\frac{3}{32}$ of an inch back of link-arc.
Distance between link-pins 14 inches.

Cut-off.			Lead.	Slip.	Travel.	Port Opening.
a H.E.	b C.E.	Diff. $a\ \&\ b$				
6	$5\frac{5}{8}$	$\frac{1}{16}$	$\frac{15}{32}$	$\frac{1}{8}$	$2\frac{1}{32}$	$\frac{15}{32}$
9	$8\frac{15}{16}$	$\frac{1}{16}$	$\frac{7}{16}$	$\frac{3}{16}$	$2\frac{9}{16}$	$\frac{17}{32}$
12	$11\frac{7}{8}$	$\frac{1}{8}$	$\frac{7}{16}$	$\frac{1}{4}$	$2\frac{23}{32}$	$\frac{5}{8}$
15	$14\frac{13}{16}$	$\frac{3}{16}$	$\frac{13}{32}$	$\frac{11}{32}$	$3\frac{1}{16}$	$\frac{3}{4}$
18	$17\frac{5}{8}$	$\frac{3}{8}$	$\frac{3}{8}$	$\frac{1}{2}$	$3\frac{11}{16}$	$1\frac{1}{32}$
21	$20\frac{3}{8}$	$\frac{5}{8}$	$\frac{9}{32}$	$\frac{7}{8}$	$4\frac{27}{32}$	$1\frac{3}{4}$
$22\frac{1}{2}$	22	$\frac{1}{2}$	$\frac{1}{16}$	$1\frac{9}{32}$	$6\frac{3}{16}$	$2\frac{21}{32}$

In this example the reverse-shaft was set square, i.e. with the arm horizontal at midgear as shown on Pl. XX, and the cut-off was equalized at $\frac{1}{8}$ stroke only. It may be considered to be the typical example. The lead at full-gear, or with the cut-off at $22\frac{1}{2}$ inches, is small, but increases rapidly as the cut-off is shortened, till it is nearly half an inch when the cut-off is at quarter-stroke; meanwhile the travel and port-opening both decrease rapidly. The travel of the valve in full-gear is more than twice the eccentricity, due to the slip of the link-block and to other irregularities.

An inequality in the cut-off of $\frac{1}{4}$ of an inch in 24 inches, i.e. one per cent of the stroke, cannot be distinguished in the running of the engine or on the indicator-diagram; consequently the cut-off may be considered to be equalized from $\frac{1}{4}$ to $\frac{5}{8}$ of the stroke. The inequality of the cut-off becomes as large as $\frac{5}{8}$ of an inch at 18 and 21 inches; but while such an inequality may possibly be distinguished on an indicator-diagram, it cannot have an appreciable effect on the running of the engine.

In all the examples the gear was made symmetrical; consequently the action in backing gear was almost identical with that in forward gear, and need not be stated separately.

TABLE II.

Rocker used—link-pins $2\frac{7}{8}$ inches back of link-arc.
Cut-off equalized at $\frac{1}{4}$ stroke.
Saddle-pin $\frac{1}{15}$ of an inch back of link-arc.
Distance between link-pins 14 inches.

Cut-off.			Lead.	Slip.	Travel.	Port Opening.
a H.E.	b C.E.	Diff. a & b.				
6	$5\frac{25}{32}$	$\frac{7}{32}$	$\frac{7}{16}$	$\frac{1}{4}$	$2\frac{7}{16}$	$\frac{15}{32}$
9	$8\frac{25}{32}$	$\frac{7}{32}$	$\frac{7}{16}$	$\frac{1}{4}$	$2\frac{17}{32}$	$\frac{17}{32}$
12	$11\frac{3}{4}$	$\frac{1}{4}$	$\frac{13}{32}$	$\frac{9}{32}$	$2\frac{11}{16}$	$\frac{5}{8}$
15	$14\frac{23}{32}$	$\frac{9}{32}$	$\frac{3}{8}$	$\frac{3}{8}$	$3\frac{1}{32}$	$\frac{25}{32}$
18	$17\frac{5}{8}$	$\frac{3}{8}$	$\frac{11}{32}$	$\frac{1}{2}$	$3\frac{9}{16}$	$1\frac{1}{32}$
21	$20\frac{5}{8}$	$\frac{3}{8}$	$\frac{1}{4}$	$\frac{25}{32}$	$4\frac{13}{16}$	$1\frac{21}{32}$
$22\frac{1}{2}$	$22\frac{5}{32}$	$\frac{11}{32}$	H.E. $\frac{1}{8}$ C.E. $\frac{1}{16}$	$1\frac{1}{8}$	$6\frac{5}{16}$	$2\frac{11}{16}$

In this example the eccentric-rods were made one inch longer than in the preceding example, which made the radius of the link-arc one inch shorter than required for the normal condition, i.e. 48 instead of 49 inches. The effect was to reduce the inequality of the cut-off at long cut-off, so that the cut-off may be considered to be practically equal for the entire stroke. The full-gear lead had an inequality of $\frac{1}{16}$ of an inch, but at other grades of the link the lead was sensibly equal. Had the link been set to give equal lead at full-gear, then the mid-gear lead would have had an inequality of $\frac{1}{16}$ of an inch in half an inch, which would not have any appreciable effect in the running of the engine.

TABLE III.

Rocker used—link-pins $2\frac{7}{8}$ inches back of link-arc.
Cut-off equalized at $\frac{1}{2}$ stroke and at full-gear.
Saddle-pin $\frac{5}{32}$ of an inch back of link-arc.
Distance between link-pins 14 inches.

Cut-off.			Lead.	Slip.	Travel.	Port Opening.
a H.E.	b C.E.	Diff. a & b.				
6	$5\frac{11}{32}$	$\frac{11}{32}$	$\frac{7}{16}$	$\frac{9}{32}$	$2\frac{18}{32}$	$\frac{7}{16}$
9	$8\frac{18}{32}$	$\frac{9}{32}$	$\frac{7}{16}$	$\frac{5}{16}$	$2\frac{1}{2}$	$\frac{1}{2}$
12	12	0	$\frac{13}{32}$	$\frac{13}{32}$	$2\frac{3}{4}$	$\frac{5}{8}$
15	$14\frac{31}{32}$	$\frac{1}{32}$	$\frac{3}{8}$	$\frac{1}{4}$	$3\frac{1}{16}$	$\frac{28}{32}$
18	$17\frac{21}{32}$	$\frac{3}{32}$	$\frac{5}{16}$	$\frac{3}{16}$	$3\frac{3}{8}$	$1\frac{1}{16}$
21	$20\frac{15}{16}$	$\frac{1}{16}$	$\frac{7}{16}$	1	$4\frac{1}{16}$	$1\frac{3}{4}$
$22\frac{1}{2}$	$22\frac{18}{32}$	$\frac{8}{32}$	H.E. 0 C.E. $\frac{1}{16}$	$1\frac{1}{4}$	$6\frac{1}{2}$	$2\frac{1}{4}$

In the third example the cut-off was equalized at $\frac{1}{2}$ stroke and at full-gear as recommended by Auchincloss in his *Link and Valve Motions*, by the method described on page 77 and shown on Pl. XXI. With the ratio of crank to connecting-rod used in these examples, i.e. 1 : 5, the reverse-shaft was found to interfere with the eccentric-rods at full-gear backing. The table shows that the equalization of the cut-off was nearly perfect from half-gear to full-gear. The inequality in the cut-off increases as the cut-off is shortened, that is, at the place where it has the most deleterious effect; but it must be admitted that an inequality of $\frac{3}{8}$ of an inch in 6 inches cannot have a very bad effect, if indeed it should be distinguishable in the running of the engine. Both the danger of interference of the reverse-shaft with the eccentric-rods and the inequality of cut-off near mid-gear will be found to be less with a more favorable ratio of crank to connecting-rod, or with the cut-off equalized at $\frac{1}{4}$ and $\frac{3}{4}$ stroke as recommended on page 77 and shown on Pl. XXI.

TABLE IV.

Rocker used—link-pins 3 inches back of link-arc.
Cut-off equalized at $\frac{1}{2}$ stroke.
Saddle-pin $\frac{23}{32}$ of an inch back of link-arc.
Distance between link-pins 12 inches.

Cut-off.			Lead.	Slip.	Travel.	Port Opening.
a H.E.	b C.E.	Diff. a & b.				
6	$5\frac{7}{8}$	$\frac{1}{8}$	$\frac{3}{8}$	$\frac{5}{8}$	$2\frac{5}{16}$	$1\frac{3}{32}$
9	$8\frac{1}{16}$	$\frac{1}{16}$	$\frac{3}{8}$	$\frac{11}{16}$	$2\frac{15}{32}$	$1\frac{5}{32}$
12	$11\frac{1}{16}$	$\frac{1}{16}$	$\frac{11}{32}$	$\frac{3}{4}$	$2\frac{5}{8}$	$\frac{9}{16}$
15	$14\frac{5}{8}$	$\frac{3}{8}$	$\frac{5}{16}$	$\frac{25}{32}$	$2\frac{15}{16}$	$\frac{3}{4}$
18	$17\frac{3}{8}$	$\frac{5}{8}$	$\frac{5}{16}$	$\frac{15}{16}$	$3\frac{1}{2}$	1
21	$20\frac{3}{8}$	$\frac{5}{8}$	$\frac{7}{32}$	$1\frac{1}{4}$	$4\frac{13}{16}$	$1\frac{5}{8}$
$22\frac{1}{2}$	22	$\frac{1}{2}$	H.E. $\frac{3}{32}$ C.E. $\frac{1}{16}$	$1\frac{3}{4}$	$6\frac{7}{16}$	$2\frac{7}{16}$

The fourth example had the reverse-shaft set square and differed from the first example in that the link was made shorter, i.e. 12 instead of 14 inches between the link-pins; the link-pins were also set back 3 inches instead of $2\frac{7}{8}$, but such a change has not much effect. A comparison of this table with Table I will show that shortening the link increases the irregularities of the link-motion, and especially that it increases the slip at all gears.

TABLE V.

Link acts directly on valve-spindle without rocker.
Link-pins on link-arc.
Cut-off equalized at ⅜ stroke.
Saddle-pin ¼ of an inch forward of link-arc.
Distance between link-pins, 12 inches.

Cut-off.			Lead.	Slip.	Travel.	Port Opening.
a H.E.	b C.E.	Diff. a & b.				
6	$6\frac{1}{8}$	$\frac{1}{8}$	$\frac{5}{8}$	$\frac{3}{16}$	$2\frac{5}{16}$	$1\frac{3}{32}$
9	$9\frac{1}{16}$	$\frac{1}{16}$	$\frac{5}{8}$	$\frac{7}{32}$	$2\frac{11}{32}$	$1\frac{5}{32}$
12	$11\frac{3}{4}$	$\frac{1}{4}$	$\frac{5}{8}$	$\frac{1}{4}$	$2\frac{19}{32}$	$1\frac{9}{16}$
15	$14\frac{3}{8}$	$\frac{5}{8}$	$\frac{11}{32}$	$\frac{5}{16}$	$2\frac{5}{8}$	$1\frac{11}{16}$
18	$17\frac{1}{8}$	$\frac{7}{8}$	$\frac{11}{32}$	$\frac{5}{8}$	$3\frac{3}{8}$	H.E. 1 C.E. $\frac{7}{8}$
21	$20\frac{3}{16}$	$1\frac{3}{16}$	$\frac{7}{32}$	$\frac{7}{8}$	$4\frac{5}{8}$	H.E. $1\frac{5}{8}$ C.E. $1\frac{1}{2}$
$22\frac{1}{2}$	$21\frac{7}{8}$	$\frac{5}{8}$	$\frac{3}{32}$	$\frac{7}{8}$	$6\frac{1}{16}$	H.E. $2\frac{7}{8}$ C.E. $2\frac{3}{16}$

This example was chosen to represent the type of link-motion which acts directly on the valve-spindle without the intervention of a rocker, and, as is customary, has the link-pins on the link-arc. It may be compared with Table IV, which has the link-pins the same distance apart. The equalization of cut-off from ¼ to ⅝ stroke is good, though not perfect, and the inequality at and near full-gear cannot have a very bad effect, though it is much larger than in the fourth example. The slip is less than for that example at all grades; but probably the slip could be much reduced in such a link-motion with a rocker, if the link-pins were placed nearer the link-arc or on the link-arc. Such an arrangement would show greater inequality in cut-off than is found in Table IV.

The sixth example was chosen to show the effect of placing the link-pins *ahead* of the link-arc, on a link which acted directly on the valve-spindle without the intervention of a rocker. Such an arrangement would require the use of a side-bar link or a box link (shown by Fig. 1, Pl. XIV, and Fig. 3, Pl. XVIII), and might involve some mechanical difficulties.

TABLE VI.

Link acts directly on valve-spindle without rocker.
Link-pins 3 inches ahead of link-arc.
Cut-off equalized at $\frac{1}{8}$ stroke.
Saddle-pin $1\frac{3}{8}$ of an inch ahead of link-arc.
Distance between link-pins, 12 inches.

Cut-off.			Lead.	Slip.	Travel.	Port Opening.
a H.E.	b C.E.	Diff. a & b				
6	$6\frac{3}{8}$	$\frac{3}{8}$	$\frac{3}{8}$	$1\frac{23}{32}$	$2\frac{9}{32}$	$\frac{3}{8}$
9	$8\frac{3}{32}$	$\frac{1}{32}$	$\frac{3}{8}$	$1\frac{11}{16}$	$2\frac{3}{8}$	$\frac{7}{16}$
12	$11\frac{15}{16}$	$\frac{1}{16}$	$\frac{11}{32}$	$1\frac{21}{32}$	$2\frac{3}{8}$	$\frac{7}{16}$
15	$14\frac{3}{4}$	$\frac{1}{4}$	$\frac{5}{16}$	$1\frac{21}{32}$	$2\frac{13}{16}$	$\frac{81}{32}$
18	$17\frac{11}{16}$	$\frac{5}{16}$	$\frac{1}{4}$	$1\frac{5}{8}$	$3\frac{1}{4}$	$\frac{7}{8}$
21	$20\frac{1}{16}$	$\frac{5}{16}$	$\frac{3}{16}$	$1\frac{5}{8}$	$4\frac{3}{4}$	$1\frac{4}{16}$
$22\frac{1}{2}$	$22\frac{3}{8}$	$\frac{1}{8}$	$\frac{1}{32}$	$1\frac{5}{8}$	$6\frac{7}{8}$	$2\frac{11}{16}$

The equalization of the cut-off must be considered to be satisfactory, but it is attained at the expense of an excessive slip of the link-block. Probably a compromise between the link-motions shown by Tables V and VI, having the link-pins an inch or an inch and a half ahead of the link-arc, would be found to give a fair equalization of the cut-off without excessive slip. Also greater distance between the link-pins (14 instead of 12 inches) could be used with advantage.

CHAPTER IV.

RADIAL VALVE-GEARS.

THE name radial valve-gear has been applied to a number of reversing-gears that differ widely in detail and in general appearance, but agree in that they derive the mid-gear motion of the valve from some source that is equivalent to an eccentric with 90° angular advance, and they combine with this motion another that is equivalent to that of an eccentric with no angular advance. The general conception of this form of valve-gear is most easily obtained from an example.

Walschaert Gear.—This gear is chosen as the first example of the type because the elements are easily distinguished. In Fig. 1, Plate XXII, H is the engine cross-head, and a is the head of the valve-spindle. The valve is moved through a radius-rod, one end of which carries a block that may be set at any position in a slotted link dF, and the other takes hold of a combining-lever af, that receives motion from the engine cross-head. The slotted link swings on a fixed trunnion at G and is moved by an eccentric OE, which has no angular advance. In Fig. 2 the diagram in thin lines shows the gear at a dead-point, and the diagram in heavy lines shows the gear when the crank has moved through the angle $C_0OC = \theta$.

If the motion of the engine cross-head can be considered to be harmonic, then it is clear that the motion that it gives to the valve could be derived from an eccentric with 90° angular advance; this motion is made equal to the lap plus the lead.

If the block d is at the middle of the link, the valve will derive motion from the cross-head only and the mechanism will be at mid-gear. The radius of the link-arc is made equal to the length de of the radius-rod, consequently the lead is constant for all settings of the gear. If the point h of the guiding-link hf were a fixed point, then the valve would receive motion from the eccentric OE, which has no angular advance. By placing the link-block nearer the trunnion G the motion is reduced; for example, the motion communicated from the eccentric OE will be half as much if the block is half-way between d and G. If the link-block is below G the motion is reversed.

The displacement from mid-position of a valve moved by an eccentric is

$$e = r \sin (\theta + \delta).$$

Now the motion derived from the cross-head is equivalent to that from an eccentric having 90° angular advance, provided the cross-head motion is assumed to be harmonic. Consequently the valve derives a displacement from this source of

$$e_1 = r_1 \sin (\theta + 90°) = r_1 \cos \theta. \quad \cdots \quad (53)$$

From the proportions of the combining-lever and the length R of the crank, we have

$$r_1 = \frac{ae}{ef} R.$$

The displacement of the valve from the influence of the eccentric OE is

$$e_2 = r_2 \sin (\theta + 0°) = r_2 \sin \theta, \quad \cdots \quad (54)$$

in which

$$r_2 = OE \frac{dG}{GF} \times \frac{af}{ef}.$$

The entire displacement e of the valve at any crank-angle is the sum of the displacements from the two independent sources.

$$\therefore e = e_1 + e_2 = r_1 \cos\theta + r_2 \sin\theta; \quad \ldots \quad (55)$$

and since r_1 and r_2 are constant for any grade of the link, equation (54) is a special case of equation (29), r_1 and r_2 being the coördinates of the diameter of a valve-circle for that grade of the gear.

In Fig. 1, Plate XXI, the valve-circle OP_0 is drawn with a diameter r_1 and represents the mid-gear action of the valve. The circles Op, Op_1, and Op_2 represent the motions derived from the eccentric OE, at full-gear and at two intermediate gears. The circles OP, OP_1, and OP_2 represent the actual displacements of the valve, derived from both sources. It is evident that

$$OP = \left\{ \overline{OP_0}^2 + \overline{Op}^2 \right\}^{\frac{1}{2}},$$

and that OP_1 and OP_2 may be obtained in a similar manner. A comparison of Fig. 1, Pl. XXI, with Fig. 3, Pl. XVII, shows that the action of the Walschaert gear is equivalent to that of the Gooch link-motion. To aid in this comparison, the dimensions OP_0 and P_0P were transferred from Fig. 3, Pl. XVII, to Fig. 1, Pl. XXI, and consequently the diagrams are identical.

As actually constructed, this gear does not give harmonic motion to the valve, for the motion of the cross-head of the engine with the usual proportions of locomotives has considerable irregularity from the angularity of the connecting-rod; also some irregularity is introduced by the combining-lever af. Consequently such a diagram as Fig. 1, Pl. XXI, can be of use only in roughly blocking out a gear. The real action of the gear can be determined either by constructing diagrams similar to Fig. 2 on as large a scale as convenient, or by aid of a model.

A combination of the two methods, similar to the skeleton model for link-motions, may be found convenient for this purpose. Since part of the motion of the valve is derived from the cross-head, the adjustment of the gear to give equal cut-off will generally be easier than for a link-motion.

In laying out a Walschaert gear, the combination-lever af should be made vertical when the cross-head is at the middle of its stroke; the guiding-link hf should be made to vibrate equal angles above and below a horizontal line; a line from G to F on the link should be made vertical when the engine is on a dead-point; and the supporting-link with the reverse-arm ST should be so laid out that it may be guided nearly on a horizontal line, unless the adjustment may be found to require a different arrangement. The length of the combining-lever should be so chosen that its angular vibration shall not exceed 60°.

The main dimensions of the gear for any engine will be imposed on the designer by the general proportions of the engine and its frame. There are, however, two elements over which the designer will have more or less control: they are the position of the axis of the trunnion G, which in the figures is on the link-arc, but which may be placed either forward of or back of the link-arc; and the reverse-shaft T, which may often be located at will, within limits. The first will be found to have the most influence on the action of the gear.

The eccentric OE is sometimes replaced by a return-crank from the engine-crank C. The link is sometimes turned the other way, in which case the radius-rod extends forward from the head of the valve-rod.

Marshall Valve-gear.—Plate XXIII shows the Marshall valve-gear as applied to the U.S.S. *Yorktown*. In Fig 1, XX' is the axis of the cylinder, O is the centre of the shaft, and C is the crank-pin. The eccentric centred at E gives motion to a short and massive eccentric-rod EG, which is guided at o by the link oD, and is connected to the valve-spindle V by a valve-rod GV. The guiding-link oD is supported by the bell-crank lever DoS, having its axis at o; the rod ST gives connection with the reverse-shaft arm TU. In the figure the engine is at a dead-point, so that the guided point of the eccentric-rod coincides in projection with the axis of the bell-crank lever.

Fig. 2 shows the centre-lines of the gear in two positions; the heavy lines are for the full-gear of the valve-motion, and the fine lines are for a gear between that and the mid-gear. OC is the centre-line of the crank; E is the centre of the eccentric; FD is the guiding link; and oD is the arm of the bell-crank lever, having its axis at o.

When the bell-crank lever is set to give the mid-gear action of the valve, D is found at D_0, and the guided point F moves on an arc of a circle that nearly coincides with the line OY; the point E describes a circle, and all other points of the eccentric-rod describe ovals that are more or less elongated as they are near or removed from the guided point F. In this setting of the gear the horizontal motion of the point G is made equal to the lap plus the lead, so that the valve receives a motion like that given by an eccentric having 90° angular advance; as G is beyond F, the eccentric E properly coincides with the crank.

At any other gear than mid-gear, for example with D set for full-gear, the vertical displacement of F will have two components, one along the axis OY, and one perpendicular to it. The second component with some modification is transferred to G and gives to the valve an additional displacement like that from an eccentric with no angular advance. This gear is consequently of the general type described at the beginning of the chapter, but the various irregularities of the gear are so marked that the valve-diagrams similar to Fig. 1, Pl. XXI, cannot be used at all in designing and laying out the gear. Since the guided point F is always brought into coincidence with the axis of the bell-crank lever when the engine is on a dead-point, the lead is the same for all gears.

The cut-off is shortened by making D approach D_0; the thin lines show the gear set for a cut-off at about $\frac{3}{8}$ stroke. The engine is reversed by carrying D beyond D_0 toward D''.

Fig. 3 gives the valve-ellipses for full-gear and for a short cut-off, corresponding with the diagrams in full lines and finer lines shown by Fig. 2. The valve-ellipses show the defect of the gear, which is a marked inequality in the maximum port-opening. In spite of this defect, the gear has been much used for horizontal marine engines, as it interferes less than would a link-motion with the access to the working parts of the engine when running.

In some cases the point G is taken between the eccentric and the guided point, in which case the eccentric is set opposite the crank.

The design of the Marshall valve-gear must be carried out by the aid of diagrams or a model, or by a combination of the two methods, which appears to be well adapted for this work.

Hackworth Valve-gear.—This gear differs from the Marshall gear in having the guided point carried by a block that slides in straight guides and thus avoids the irregularity due to the guiding-link. The irregularity of the valve-motion is less than when the Marshall gear is used, and the maximum port-openings can be made nearly equal. The pressure on the sliding-block is large, especially at full-gear, and unless ample wearing surface is provided the friction and wear are liable to be excessive. In some cases the sliding block has been provided with rollers to reduce the friction.

Joy Valve-gear.—An example of the Joy valve-gear used on the Pennsylvania Railroad Company's tugboat *Delaware* is shown by Pl. XXIV. XX' is the centre-line of the crank and connecting-rod, and xx' is the centre-line of the valve-spindle. The lever abc is guided on a flat arc by the rod gc, and is attached to the connecting-rod at the point a, which describes an oval having the length $a_1 a_2$ equal to the stroke of the engine. This oval, which is omitted to avoid confusion of the diagram, is symmetrical with regard to the axis XX', and is slightly more pointed at the cross-head end than at the crank end. The point b, which describes the irregular oval bb_1b_2, takes the place of the centre of the single eccentric used with the Marshall valve-gear (Pl. XXIII), and acts on the lever bie. The point i of the lever bie is guided on the circular arc ff_1 by the sliding-block B, and the point e, which describes the oval ee_1e_2, carries the valve-rod ed. The connecting-rod CD, the valve-rod ed, and the rod cg are in one plane; the levers ac and be and the curved guide-bar ff_1 are double, one system being on each side of the connecting-rod; in the figure the system of levers in front of the connecting-rod is omitted to show the construction more clearly. The guided point i could evidently be guided on the arc ff_1 by a link centred at k; such a construction is frequently used in marine engines. A comparison of this gear with the Marshall valve-gear will show much similarity. The essential points of difference are: (1) the radius of the guiding-arc ff_1 is always equal to the length of the valve-rod, and (2) the irregularity due to the angularity of the lever bie is compensated by the action of the lever ac, somewhat in the manner that the linkage known as Watt's parallel motion is made to give nearly a straight-line motion. These advantages are attained at the expense of greater complication and cumbersomeness; in passing it may be remarked that the inequality of port-opening, which is the notable defect of the Marshall gear, may be nearly if not quite remedied by making the length of the guiding-link equal to the valve-rod, but such a construction is usually impracticable since it requires either an impossible length for the guiding-link, or else a short valve-rod and a long valve-spindle that must be guided at the outer end. The guiding-bars ff_1 are hung on trunnions with the axis at the point i_1, and are connected at f to the reversing-lever. The gear is shown at full-gear for left-handed rotation; it may give a shorter cut-off if the guiding-bars ff_1 are given less inclination from the horizontal or mid-gear position, and when in mid-gear it will give the valve a motion equal to the lap plus the lead; if the guiding-bars ff_1 are inclined the other way the engine will be reversed.

This gear, when properly proportioned, gives a rapid motion to the valve when opening and closing, less compression at short cut-off than does a link-motion, and the cut-off can be made nearly equal for all grades of the gear. Like all other radial valve-gears, it gives a constant lead. Its defects are, the number of parts and of joints that are liable to wear loose, and the obstruction that it offers to inspection and care of the crank-pin and cross-head when the engine is running.

To lay out a Joy valve-gear: choose a point a on the connecting-rod, having a transverse throw equal to twice the maximum displacement of the valve; make the length of the lever ac such that the angle $a_1 c_1 a_2$ shall not be more than 90°; draw the centre-line xx' of the valve-gear, and locate the point e_0 opposite the middle-point of the line $a_1 a_2$; lay off $e_0 e_1 = e_0 e_2$ equal to the lap plus the lead; lay off $a_1 b_1 = a_2 b_2$ equal to about one third of $a_1 c_1$; and draw the lines $e_1 b_1$ and $e_2 b_2$ intersecting at i_1: this last point locates the axis of the trunnions carrying the guiding-bars ff_1. If the point i is guided by a link, then the arm carrying the link must be centred at i_1; in the figure the position of such an arm is shown by the line $i_1 k$. Finally, the valve-ellipse should be drawn for several grades of the gear; in the figure the valve-ellipse oo_1 has its length equal to the stroke of the engine, and the valve-displacements are magnified fourfold, i.e. $st = 4d_0 d$. Usually the axis xx' of the valve-motion is determined by the general design, and cannot be changed much, if at all. The length of the lever ac will commonly be as great as desirable if the angle $a_1 c_1 b_1$ is something less than 90°; it cannot be made shorter without throwing excessive stress on the links and levers. The transverse motion of the point a is properly twice the maximum displacement of the valve, in order that the inclination of the guide-bars ff_1 may not be more than 25° or 30°; should such a location bring the lever ac too near the shaft, as may be the case when the crank is counterweighted, then a may be placed nearer the cross-head, but at the expense of more inclination of the guiding-bars at long cut-off. The location of the point b is under the control of the designer, and may be used to equalize the cut-off either at that grade at which the engine is to run habitually, or else to give nearly equal cut-off at all grades. The equalization of the cut-off must be made by trial, and no general rule can be given, since the elements, such as the length of the lever ac and the distance between the axes xx' and XX', over which the designer has little control, have a large influence. A skeleton model may be used to advantage in this work. It should have rods to represent the connecting-rod, the levers ac and be, and the links cg and ki. The point k will be located on an arc of a circle centred at i_1. The points i and b on the levers be and ac should be made adjustable; the first by mounting it on a sliding-block that can be clamped in any desired position, and the second by that method, or by a series of holes to receive the screw representing the pin at b. It may be found advantageous to provide pieces, properly guided, to represent the cross-head and head of the valve-spindle, but a simple model may be made by placing these points by hand on the lines XX' and xx' for the several positions of the model, and the centre C may in like manner be placed on the circle $CC''C'$.

CHAPTER V.

DOUBLE VALVE-GEARS.

A PLAIN slide-valve, set to give an early cut-off, is liable to give either an excessive compression or an early release, or both. A single valve under the control of a gear that gives a variable cut-off, such as a shifting-eccentric or a link-motion, is open to the same difficulties; and in addition the compression varies with the cut-off, though to a less degree. For a stationary engine a large compression may be undesirable, and a varying compression is always so. To avoid these difficulties two valves are frequently used; one, called the main valve, has an unvariable motion, and gives the admission, release, and compression; the other, called the cut-off valve, gives the cut-off only, which may be varied without affecting the other events of the stroke.

The cut-off valve may be placed in a separate valve-chest, as shown by Fig. 3, Pl. XXV, or it may be placed on the back of the main valve, as shown by Fig. 1, Pl. XXVI; thus giving rise to two separate types of double valve-gears. It is important to obtain a clear conception of the principles of double valves, and then all existing forms of double valve-gears may be readily understood, and a gear for a given purpose may be easily designed, or else it may be shown that a satisfactory design is impossible.

Cut-off Valve in a Separate Valve-chest.—The usual arrangement of this valve-gear is shown by Fig. 3, Pl. XXV. The main valve, which receives motion from an eccentric with constant angular advance and eccentricity, is designed to give the desired release and compression, and is set to give equal lead; it will be observed that either the release or the compression may be equalized. In the figure there is no inside lap; this arrangement may be frequently found desirable, but it is chosen here for the sake of simplicity, and will be adhered to throughout the chapter; attention will be given exclusively to the cut-off, since the other features of the gear are the same as for a plain slide-valve and have received sufficient attention in the first chapter.

In the figure the valves are shown disconnected from the eccentrics and both in mid-position; they cannot both be in such position when the gear is connected up, but such a drawing is convenient in laying out the valves. The cut-off valve is a rectangular open frame having the acting edges inside. The distance l from one edge of the valve to the opposite edge of the port is the *clearance* of the valve; when the valve is displaced from mid-position an amount equal to the clearance, the cut-off valve gives either *cut-off* or *readmission*. The righthand edge gives cut-off for the head end of the engine, and readmission for the crank end; it is important that the readmission by the cut-off valve should precede the admission by the main valve, in order that the second steam-chest shall then be properly filled with steam at full pressure.

The steam-port a_0 is for the passage of full-pressure steam only, and may consequently be made from $\frac{1}{2}$ to $\frac{2}{3}$ of the area of the port a, through which exhaust steam also must pass. Two or more ports are frequently provided in the cut-off valve-seat, and the valve is then known as a gridiron valve; in such case the combined area of all the ports a_0 should be a little in excess of what would be given to one port, to allow for the greater friction in numerous narrow passages. A gridiron valve acting on several narrow ports will require a proportionately less throw.

In Fig 1, Pl. XXV, OP is the diameter of the valve-circle

for the main valve which gives admission at OR_0 and $OR_0{}'$, and cut-off at OR. The eccentric acting on the cut-off valve is given a negative angular advance, i.e. it is less than 90° in advance of the crank; consequently the displacements of the cut-off valve from mid-position are given by a valve-circle, such as OP_0, having its diameter laid off at an angle δ_0 (the negative angular advance) *away* from the crank. If the clearance is equal to Ol_2, then the cut-off by the cut-off valve occurs at OR_2 and the readmission at OR.

A variable cut-off by aid of this gear may be obtained by varying the clearance of the valve, or the throw of the eccentric, or by using a shifting-eccentric. Now the same effect is produced by increasing the lap (or decreasing the clearance) with a constant eccentricity as is produced by decreasing the eccentricity with a constant lap; consequently an investigation of one of the two methods will serve for both. It will appear that neither will give a good variable cut-off. On the other hand, a satisfactory gear may be had by using a shifting-eccentric.

In Fig. 1, Pl. XXV, the clearance Ol_2 gives cut-off at OR_2 and readmission at OR, coincident with the cut-off by the main valve: and that is the latest admissible cut-off; for if the clearance is made equal to Ol_1 in order to obtain a cut-off at OR_1, the readmission will occur at $OR_1{}'$, before the main-valve cut-off, and a double admission of steam will occur. Such a double admission causes a large waste of steam and an irregular action of the engine and cannot be tolerated. The earliest admissible cut-off is obtained by a clearance equal to Ol_4, which gives cut-off at OR_4 and readmission at $OR_4{}'$. An inspection of the figure shows that the angle R_1OR_2 must, from symmetry, be equal to $ROR_0{}'$; the latter angle depends on the lap of the main valve, and it is at once evident that only a limited range of cut-off can be obtained with such an arrangement.

In practice a gridiron cut-off valve was commonly used and the cut-off was obtained by varying the travel of the valve.

For this purpose a fixed eccentric was connected to the end of a slotted lever or *link*, and motion was communicated to the valve from a link-block that could be set at any desired place in the link. To shorten the cut-off, the link-block was moved toward the fixed end or fulcrum of the link. An idea of this arrangement may be obtained by supposing the radius-rod of the Walschaert gear (Plate XXII) to act on the cut-off valve-spindle; the eccentric should, of course, have a negative angular advance, and only half of the link will be required. The effect of such an arrangement is the same as though the eccentricity were varied, and, just as has been shown to be the case for a gear with varying lap, the range of variation of cut-off depends on the lap of the main valve.

Cut-off Valve with Shifting-eccentric. — If the cut-off valve in a separate valve-chest is moved by a properly designed shifting-eccentric, the readmission may be kept within the proper limits, i.e. before the admission and after the cut-off by the main valve, and at the same time any desired range of cut-off may be had.

In Fig. 2, Pl. XXV, let OP be the main valve-circle, giving cut-off at OR and admission at OR_0'. Suppose that the cut-off is to vary from OR_1, corresponding with $\frac{1}{8}$ stroke, to OR coincident with the cut-off by the main valve. Bisect the angle ROR_0' by the line OP_0', and bisect the angle ROR_1 by the line OP_0; then a valve-circle centred at C_0' can be made to give a cut-off coincident with that by the main valve, and readmission coincident with the main-valve admission; while a valve-circle centred at C_0 can give cut-off at OR_1, and readmission coincident with the main-valve cut-off at OR.

It is convenient to have the shifting-eccentric on an arm centred on the centre-line of the crank, or rather on that line produced, for then the cut-off gear will work equally well in forward and in backing gear on a reversing-engine; but for a stationary engine such an arrangement is not essential. We now have three elements of which one may be chosen at pleasure and the other two will then be determinate. In the figure the diameter OP_0 of the valve-circle to give early cut-off is made $1\frac{1}{2}$ inches, equal to the eccentricity for the main eccentric; the clearance Ol of the cut-off valve is then determined by the intersection of that circle by the lines OR and OR_1, and is $\frac{13}{16}$ of an inch. The centre C_0' of the valve-circle to give the cut-off coincident with that of the main valve must be so chosen that it shall pass through O and l''; it is on a perpendicular to Ol'' at its middle point. Draw AS perpendicular to the middle point of an imaginary line joining P_0 and P_0'; then S, the intersection of this line with the axis XX', is the centre of an arc on which the point P_0 will travel as the cut-off is shortened. The shifting-eccentric must be swung from a point on the centre-line of the crank produced, and at a distance from the centre of the shaft equal to OS ($1\frac{1}{8}$ of an inch); when the crank is at OX this point will be at T. With ordinary proportions for an engine, the point T is liable to fall inside the shaft, or else so near that the construction of an arm for the eccentric will be impossible. A reversing-engine may have the following arrangement: fast to the shaft may be an eccentric with an eccentricity equal to OS and with its centre opposite the crank-pin; this eccentric may carry another having an eccentricity equal to AS ($1\frac{31}{32}$ of an inch); the second or outside eccentric will be turned toward the crank so that its angular advance will be negative. A stationary engine that always runs in one direction may have the swinging arm, for the eccentric controlling the cut-off valve, centred at any convenient point on the line AS produced.

There now remains the determination of the width of the cut-off valve to prevent leakage past the outside edge. The greatest displacement of the valve from mid-position is equal to $OP_0 = 1\frac{1}{2}$ of an inch; consequently the distance in Fig. 3, Pl. XXV, from the outside edge of the valve to the nearest edge of the port should be somewhat greater than $1\frac{1}{2}$ of an inch; it is made $1\frac{7}{16}$ of an inch.

Cut-off Valve on back of Main Valve.—Fig. 1, Pl. XXVI, shows a cut-off valve on the back of the main valve, both being disconnected from their eccentrics and placed in mid-position. When the gear is connected up, such a position of both valves at the same time does not occur, but it is convenient to make a drawing of the valves in that position to show the laps and other dimensions of the main valve, and the clearance and length of the cut-off valve. The main valve is designed to give the desired compression and release, and is set to give equal lead; either the compression or the release may be equalized. The cut-off valve is connected to an eccentric having a large angular advance, so that it is nearly (sometimes exactly) opposite the crank.

In Fig. 3, OP is the diameter of the main valve-circle, and OP_0 is the diameter of a valve-circle showing the absolute displacements of the cut-off valve. At any crank-position, such as OR', the chord Oc intercepted by the main valve-circle shows the displacement of the main valve; represented in Fig. 2 by e. The absolute displacement of the cut-off valve is shown by the chord Ob, intercepted by the cut-off valve-circle; represented by e_0 in Fig. 2. Both of these displacements are toward the left, but the former being the greater, the relative displacement e_x of the cut-off valve with regard to the main valve, and measured from the centre of the main valve, is towards the right, tending to shut the port in the main valve.

Now the displacement of the main valve, with an eccentricity r and an angular advance δ, is, for the crank-angle θ,

$$e = r \sin(\delta + \theta) = r \sin \delta \cos \theta + r \cos \delta \sin \theta. \quad (56)$$

The cut-off valve-eccentric has the angular advance δ_0 and the eccentricity r_0, and for the crank-angle θ the displacement of the cut-off valve is

$$e_0 = r_0 \sin(\delta_0 + \theta) = r_0 \sin \delta_0 \cos \theta + r_0 \cos \delta_0 \sin \theta. \quad (57)$$

The relative displacement of the cut-off valve measured from the middle of the main valve is

$$e_x = e - e_0 = (r \sin \delta - r_0 \sin \delta_0) \cos \theta + (r \cos \delta - r_0 \cos \delta_0) \sin \theta. \quad (58)$$

which may be written

$$e_x = A \cos \theta + B \sin \theta. \quad . \quad . \quad . \quad (59)$$

Now it has been shown on page 50 that any valve-motion that can be represented by an equation having the form of equation (59) is harmonic; and can be represented by a valve-circle, having for the coördinates of the end of the diameter of the valve-circle A and B.

Since $r_0 \cos \delta_0$ is longer than $r \cos \delta$, A will in this case be negative and must be laid off to the left of the origin. Consequently the circle representing the relative motion of the cut-off valve may be located in Fig. 3 by laying off $Ov = A$ and $vP_x = B$, and then by drawing the diameter OP_x on which the circle is to be drawn. This circle is called the *auxiliary* circle; it is to be borne in mind that the angle YOP_x is not an angular advance, nor is OP_x an eccentricity.

Auxiliary Valve-circle.—Although the auxiliary valve-circle can always be drawn by the process just stated, the usual and convenient method is as follows: In Fig. 4, Pl. XXVI, let OP and OP_0 be the diameters of the valve-circles for the main valve and the cut-off valve; with P as a centre and with a radius equal to OP_0, and with O as a centre and with a radius equal to PP_0, draw arcs intersecting at P_x; then OP_x is the diameter of the auxiliary circle. The figure P_xPP_0O is of course a parallelogram, and the process just described will be called *completing the parallelogram.*

To prove the method just given: Draw P_xv and Pu parallel to OY, and draw Ps and P_0t parallel to OX. It is evident that

$$Ov = P_0u = P_0t - Ps = r_0 \sin \delta_0 - r \sin \delta = -A ;$$
$$P_xv = Pu = Os - Ot = r \cos \delta - r_0 \cos \delta_0 = B.$$

When the displacement of the cut-off valve from the middle of the main valve is equal to the clearance l, Fig. 1, Pl. XXVI, the edge of the valve coincides with the edge of the port and we have either cut-off or readmission. The crank-position at cut-off and readmission can be found by drawing the circle $l_1 l_2$ with a radius Ol equal to the clearance l. In Fig. 3 cut-off occurs at the crank-position OR_0, corresponding to a piston-displacement xa_0 for harmonic motion; the valves are then in the position shown by Fig. 6. Readmission occurs at OR''; the edge of the cut-off valve is on the edge of the port as in Fig. 6, but the valve is then moving towards the left to open the port through the main valve; the main valve is not in the position shown by that figure. It is important to know when cut-off by the main valve occurs; in Fig. 3 it occurs at OR, corresponding, with harmonic motion, to a piston-displacement xa; the valves are then in the position shown by Fig. 7.

The readmission by the cut-off valve must not occur before cut-off by the main valve, otherwise a double admission of steam will take place. It should occur before the admission by the right-hand end of the main valve; i.e., before the crank comes to the position OR_a, Fig. 3. The readmission for the left-hand end of the main valve is of course given by the left-hand edge of the cut-off valve.

Meyer Valve.—A double valve-gear, known as the Meyer valve, is shown by Fig. 2, Pl. XXVIII. The cut-off valve is made in two parts on a valve-spindle with a right and left screw, so that the position of the plates may be adjusted by rotating the valve-spindle; thus the clearance may be changed, and consequently the cut-off may be varied. In order that this may be done while the engine is running, there is a swivel-joint in the valve-spindle between the valve-rod head and the valve-chest, and the tail of the valve-spindle is carried through the head end of the valve-chest, where it reciprocates through a hand-wheel as shown by Fig. 1; the valve-spindle is squared so that it may be rotated by turning the hand-wheel.

In Fig. 1, Pl. XXVII, the valve-circle OP, the auxiliary circle OP_x, and the diameter OP_0 for the valve-circle showing the absolute motion of the cut-off valve, are transferred from Fig. 3, Pl. XXVI. The cut-off by the main valve occurs, as in that figure, at OR. With the same clearance Ol, the cut-off by the cut-off valve is at the crank-position OR_0, 90° from the dead-point. If the clearance is increased to the amount Ol_1, the cut-off is delayed and occurs at the crank-position OR_1. With this clearance the readmission occurs at OR, coincident with the cut-off by the main valve, and a later cut-off would be accompanied by a readmission; consequently the latest admissible cut-off is at the crank-position OR_1.

In order that cut-off may occur at a given crank-position, for example at OR_2 corresponding to a piston-displacement equal to Xa_2, the clearance must be made equal to the chord $Ol_2' = Ol_2$, cut from the line OR_2 by the auxiliary circle. If the clearance becomes zero, so that in mid-position the edge of the cut-off valve coincides with the outer edge of the port in the main valve, then the cut-off comes at a crank-position OR', perpendicular to the diameter OP_x of the auxiliary circle. For an earlier cut-off, the line of the crank, for example OR_3, will cut the auxiliary circle at a point beyond the origin O; and the cut-off valve will then have a lap equal to $Ol_3' = Ol_3$ when in mid-position. For a cut-off at the dead-point a lap equal to Ol_4 will be required. The crank-position at readmission is at the second intersection of the auxiliary circle by the clearance-circle; for example, with a clearance equal to Ol_1, the readmission occurs at OR. When the valve has a lap in mid-position, for example Ol_3, the readmission occurs at a crank-position found by drawing a line from the second intersection of the clearance-circle and the auxiliary circle, towards the origin O and thence to the crank-pin circle. The readmission usually comes in the second or third quadrant, and so long as it is later than the cut-off by the main valve, it is of little importance to know just where it occurs.

Design of a Meyer Valve.—The main valve shown by Fig. 2, Pl. XXVIII, has an outside lap of half an inch, and is moved by an eccentric with an eccentricity of $1\frac{1}{2}$ of an inch; the inside lap is zero. It is set with $\frac{1}{32}$ of an inch lead. With these dimensions the valve-circle OP in Fig. 3 can be drawn, and the cut-off will be found to occur at OR_c; corresponding, with harmonic motion, to 0.89 of the stroke of the piston.

The steam-port in the valve-seat is $\frac{3}{4}$ of an inch wide, and the steam-port through the main valve may be taken to be $\frac{2}{3}$ as much, or $\frac{1}{2}$ of an inch. The diameter of the auxiliary circle may be assumed to be one inch. With an eccentricity of $1\frac{3}{4}$ of an inch for the cut-off valve-eccentric, the parallelogram PP_0OP_x may be drawn locating both the auxiliary circle OP_x and the diameter OP_0 of the circle showing the absolute displacements of the cut-off valve. The cut-off valve-eccentric has the angular advance YOP_0, equal to $55°$; it is convenient to know this angle approximately in setting the valves. The dimensions for OP_x and OP_0 are chosen by trial to give a convenient location of the auxiliary circle, with its diameter placed *beyond* OR_c. Were the auxiliary circle placed with its diameter coincident with OR_c, as shown by Fig. 5, Pl. XXVI, then a readmission would be impossible; such a disposition is recommended by Zeuner, but it has the disadvantage that the motion of the cut-off valve is very slow when giving a long cut-off. This defect is mitigated by placing the auxiliary circle beyond OR_c, as in Fig. 3, Pl. XXVIII; the earliest possible readmission is clearly at OR_2.

The largest clearance is $Ol = Ol' = \frac{21}{32}$ of an inch; the least clearance, or the greatest lap, will depend on the earliest required cut-off. Let it be assumed that the earliest cut-off shall be at OR_1, corresponding to a piston-displacement $Xa = \frac{1}{8}$ of the stroke, for harmonic motion; then the cut-off valve must have a *lap* equal to $Ol_1 = Ol_1' = \frac{7}{16}$ of an inch, nearly.

The lower face of the main valve is laid out as for a plain slide-valve, but with the additions demanded by the passage through it. The least width of bridge is equal to the eccentricity less the sum of the lap and the width of port, or $1\frac{1}{2} - (\frac{1}{2} + \frac{3}{4})$, $= \frac{1}{4}$ of an inch; the width used is $\frac{1}{2}$ of an inch. In like manner the width of the exhaust-space is $1\frac{1}{2} + \frac{3}{4} - \frac{1}{2} = 1\frac{3}{4}$ of an inch; the width used is $2 \times \frac{15}{16} = 1\frac{7}{8}$ of an inch. In order that the edge c of the passage through the valve may not reduce the passage through the port to less than $\frac{1}{2}$ of an inch, the distance ac is made $1\frac{1}{2} + \frac{1}{2} = 2$ inches; this feature is frequently overlooked and dc is carelessly made equal to ef. In order that the edge g of the valve shall not come to the edge b of the port, the distance bg is made $1\frac{1}{2} + \frac{1}{8} = 1\frac{5}{8}$ of an inch. The valve-face is cut away at a point h, $1\frac{3}{4}$ of an inch from g, thereby giving an overtravel of $\frac{1}{4}$ of an inch. In order that the space $efdc$ may be made small, the height of the exhaust-space is made only $1\frac{7}{16}$ of an inch, a dimension that is probably too small to give a perfectly free exhaust.

The width of the cut-off valve must be enough so that steam cannot leak past the inside edge when the valve is set to give the earliest cut-off and also has its maximum displacement. In Fig. 2, Pl. XXVIII, the position of the valve to give cut-off at $\frac{1}{3}$ of the stroke is shown by dotted lines; its *lap* is $\frac{7}{16}$ of an inch, and its left-hand edge k is $1 + \frac{1}{8} = 1\frac{1}{8}$ of an inch from e. The length of the cut-off valve is $1\frac{1}{8} + \frac{1}{2} + \frac{7}{16} = 2\frac{1}{16}$ inches. The cut-off valve is shown in section with a clearance of $\frac{31}{32}$ of an inch, which is proper for giving the longest cut-off coincident with that of the main valve; its left-hand edge is $2\frac{1}{16} + \frac{31}{32} = 3\frac{1}{32}$ inches from the edge f of the port ef; the distance of the edge f of the port from the middle of the valve is made $3\frac{1}{16}$ inches. The half-length of the main valve, over all, is made $4\frac{1}{4}$ inches, and provides an overtravel of $\frac{1}{4}$ of an inch for the cut-off valve when set to give the earliest cut-off.

The valve-spindle is provided with a right-and-left-hand screw, of which the right-hand part is shown. The thread should be cut only far enough to give the desired variation of cut-off, or some other stop should be provided in order that the engine attendants may not move the cut-off valve too far out, and so get a leakage or even admission of steam past the inner edge. The spindle is shown in two parts, joined by a right-and-left screw and circular nut or sleeve with pins to prevent the joint from jarring loose; this arrangement is to facilitate the erection of the valve-gear.

In this design the inside lap is made zero and the compression and release are neglected; in practice these features should receive the same attention as is accorded to them in designing a plain slide-valve. Again, the irregularity of the piston-motion due to the angularity of the connecting-rod has been ignored, and the clearance (or lap) of the cut-off valve has been made the same at both ends. This method is commonly followed in practice, but by using proper pitches for the threads on the valve-spindle the cut-off may be equalized at two points of the stroke, for example at $\frac{1}{4}$ and at $\frac{1}{2}$ stroke, and will then be found to be more nearly equal for all parts of the stroke, except for long cut-off, when inequality is of less importance.

Meyer Valve with Cut-off at Inside Edge.—Sometimes the Meyer valve is designed to cut off at the inside edge, as shown by Fig. 4, Pl. XXVIII. It is then convenient to consider that the valve has a lap ab which diminishes as the cut-off is lengthened, and which may become zero and finally change to a clearance, shown by ac when the valve is in the position indicated by dotted lines. The eccentric is given a negative angular advance, i.e. it is set somewhat less than 90° ahead of the crank.

In Fig. 2, Pl. XXVII, the main valve-circle is OP, giving a cut-off at OR, with a lap $On = On''$. Let it be assumed that the earliest required cut-off is at OR_2, and that the *latest* readmission must be at OR_1; then the auxiliary circle may have its diameter at OP_x, on a line bisecting the angle R_1OR_2. The diameter of the valve-circle for showing the absolute displacement of the cut-off valve will be found at OP_0 by completing the parallelogram PP_xOP_0; the eccentricity for the cut-off valve-eccentric is OP_0, and the negative angular advance is YOP_0. The auxiliary circle may be placed lower down, thereby giving an earlier readmission and at the same time a longer eccentricity, but it cannot be placed higher up. There is no danger of a double admission of steam at a long cut-off, as was found to be the case with the ordinary form of Meyer valve.

Meyer Valve for Reversing-engines.—If a Meyer valve is designed to give a satisfactory action in forward gear for a reversing-engine, the backing gear is liable to be very unfavorable. In Fig. 5, Pl. XXVI, the main valve-circle for forward gear is OP, and with a lap-circle $nn'n''$ the cut-off comes at OR_c, corresponding to $\frac{3}{4}$ stroke with harmonic motion. The auxiliary valve-circle has its diameter coincident with OR_c to avoid the possibility of a double admission of steam; the eccentricity is found to be equal to OP_0, and the angular advance is equal to YOP_0. In backing gear the main valve-circle is at OP', and cut-off by the main valve is at OR_c', corresponding to $\frac{3}{4}$ stroke, but the auxiliary valve-circle is now OP_x'', found by completing the parallelogram $P'P_0OP_x''$. With a clearance equal to $Ol = Ol' = Ol''$, the cut-off in forward gear occurs at half-stroke, but in backing gear the cut-off is at OR_2, corresponding to $\frac{1}{8}$ stroke. Such an arrangement cannot be used, for with it the engine cannot run at full power in backing gear. The action of the cut-off valve may be made the same in forward and backing gears by giving its eccentric 90° angular advance. In Fig. 5 the parallelogram PP'_0OP_x' is drawn with PP_0 parallel to OR_c, and with PP_x' parallel to XX', thus giving the diameter of the auxiliary circle coincident with OR_c and giving 90° angular advance, but both the cut-off eccentricity and the relative travel of the valve are thereby made excessive. In practice the cut-off eccentric is often given the same eccentricity as the main valve-eccentric, and then with 90° angular advance it is liable to give a double admission at long cut-off.

Cut-off Valve with Loose Eccentric.—Let the cut-off valve receive motion from an eccentric which may turn freely on the engine-shaft and which is under the control of a shaft-governor; let the clearance of the cut-off valve be unalterable: then the cut-off can be varied by changing the angular advance of the cut-off eccentric.

In Fig. 3, Pl. XXVII, the main valve-circle is OP, and with a lap $on = on'$ the cut-off by the main valve occurs at OR_c. The cut-off eccentric may have its angular advance changed from YOP_0 to YOP_0', and the auxiliary circle may change from OP_x to OP_x'. When the position of the diameter OP_0 of the circle showing the absolute displacement of the cut-off valve is known, the auxiliary circle may be located by completing the parallelogram PP_0OP_x; or the centre C_x of the auxiliary circle may be located by completing a parallelogram CC_0OC_x on the half-diameters OC and OC_0 of the main valve-circle and the cut-off valve-circle. Since the side CC_x of this last parallelogram is equal to $OC_0 = \frac{1}{2}OP_0$, it is at once apparent that the locus of the centre of the auxiliary circle is the dotted circle C_xC_x' drawn from the centre C of the main valve-circle, and with a radius equal to half the eccentricity of the cut-off eccentric. Again, since PP_x is equal to OP_0, the locus of the end of the diameter of the auxiliary circle is a circle drawn from P as a centre and with a radius equal to the eccentricity of the cut-off eccentric. The locus of the centre of the auxiliary circle is the more convenient for use in solution of problems.

Let it be assumed that the cut-off shall vary from the crank-position OR_c, coincident with the cut-off by the main valve, to OR_1, corresponding to $\frac{1}{4}$ stroke for harmonic motion. Assume the clearance of the cut-off valve and draw the circle $ll'l''$. Erect a perpendicular SC_x at the middle of the line Ol''; it will intersect the locus C_xC_x' at C_x, the centre of the auxiliary circle that will give a cut-off by the cut-off valve coincident with the cut-off by the main valve. In like manner, erect a perpendicular at the middle of the line Ol'; it will locate the centre C_x' of the auxiliary circle that gives a cut-off at OR_1, corresponding to ¼ stroke. In designing a valve-gear of this type, the clearance of the cut-off valve may be chosen, usually somewhat larger than the width of the port in the main valve; and then the auxiliary circle for maximum cut-off may be given such a diameter that a satisfactory action may be had in that gear. The eccentricity of the cut-off eccentric will be found by completing the parallelogram in the usual way; should the result be an undesirable dimension it may be modified, since the diameter of the auxiliary circle may be varied to a considerable extent. Finally, the auxiliary circle to give the earliest cut-off may be found by the process just stated; it is liable to have a large diameter, and the travel and wear of the cut-off valve is likely to be excessive. In the figure the auxiliary circle which gives a cut-off at ¼ stroke is one third larger than the main valve-circle, and it would be still larger for a shorter cut-off. The maximum diameter of the auxiliary circle is equal to the sum of the eccentricities for the two eccentrics.

If this gear is used with a shaft-governor, the cut-off by the main valve will commonly be earlier than that shown in Fig. 3—a circumstance that will make the design of a satisfactory gear easier. Moreover it may be possible to limit the maximum cut-off to half-stroke or less, even though the main valve gives a cut-off beyond half-stroke; in that case the valve mechanism must be so arranged that the cut-off valve cannot act beyond the assumed range of cut-off, otherwise a double admission may occur. These observations are applicable also to the next type of valve-gear.

Cut-off Valve with Constant Travel.—It is desirable that a valve shall overtravel its seat in order that the seat and the face of the valve may wear evenly and remain true. This is seldom possible for a Meyer valve with the common proportions, or for the cut-off valve under the control of a loose eccentric. It has been seen that the design of the cut-off valve is conveniently begun by choosing the diameter and position of the auxiliary circle; it will be found that the design of a double valve-gear for a given purpose may be worked out by first finding how the auxiliary circle must be located or changed to give the desired action, and then finding how the cut-off eccentric must move to produce such an auxiliary circle.

Suppose that the auxiliary valve-circle is to have a constant diameter, and that the variation in cut-off is to be produced by swinging the auxiliary circle around the origin O, Fig. 1, Pl. XXIX, from the position OP_x to OP_x'. With a clearance equal to $Ol = Ol' = Ol''$, the first-named auxiliary circle will give cut-off at OR_c, coincident with the cut-off by the main valve; and the other auxiliary circle, OP_x', will give cut-off at OR_1, corresponding to $\frac{1}{8}$ stroke. The diameters of the cut-off valve-circles, showing absolute displacements, are OP_0 and OP_0', found by completing the parallelograms PP_xOP_0 and $PP_x'OP_0'$. It is evident that the locus of the point P_0 is the circle P_0P_0' drawn from P as a centre and with a radius equal to the diameter of the auxiliary circle.

The arrangement of the eccentrics for this type of valve-gear is shown by Fig. 2, Pl. XXIX. The centre of the engine-shaft is at O; on the shaft is the fixed eccentric centred at E, for giving motion to the main valve; the cut-off eccentric is carried by the main eccentric, and is shown by the full-line circle with its centre at E_0', corresponding to OP_0' in Fig. 1, while the dotted circle shows it with the centre at E_0, corresponding to OP_0 in Fig. 1. The cut-off eccentric may readily be placed under the control of a shaft-governor.

The main valve represented by Fig. 3, Pl. XXIX, has a lap of $\frac{1}{2}$ of an inch, and is moved by an eccentric having $1\frac{1}{2}$ of an inch eccentricity, and is consequently a reduplication of the main valve for the Meyer valve-gear shown on Pl. XXVIII, at its lower surface; the top is of course laid out after the cut-off valve has been designed. The main valve-circle has its diameter at OP, Fig. 1, and the cut-off by that valve occurs at OR_c. The cut-off valve is a double-ported or gridiron valve, each of the ports being $\frac{1}{4}$ of an inch wide. The clearance of the cut-off valve is assumed to be $\frac{3}{8}$ of an inch, represented by the circle $ll'l''$. As the auxiliary valve swings toward the right to give an earlier cut-off, the readmission moves through the same angle toward the line OR_c, the crank-position at cut-off by the main valve; and it is at once evident that the readmission must not be earlier than OR_c, otherwise a double admission may occur. The smallest admissible auxiliary circle will have its centre on the line Os bisecting the angle R_1OR_c, and it will pass through the points l' and l'' at the intersection of the clearance-circle by the lines OR_1 and OR_c. The diameter chosen for the auxiliary circle is $\frac{3}{4}$ of an inch, or twice the clearance of the cut-off valve. The extreme positions, OP_x' and OP_x, of the auxiliary circle are so located that they shall pass, one through l' and the other through l''; they give cut-off at OR_1 and at OR_c. The corresponding diameters of the cut-off valve-circles are OP_0' and OP_0, found by completing the parallelograms P_xPP_0O and $P_x'PP_0'O$. The cut-off eccentric is mounted on the main eccentric and has an eccentricity, referred to that eccentric, of $\frac{3}{4}$ of an inch, equal to the diameter of the auxiliary circle.

In laying out the cut-off valve and the upper surface of the main valve, it is convenient to begin in Fig. 3 by placing the port a in a convenient position near the exhaust-space e. From the right-hand edge of this port lay off $\frac{3}{4}$ of an inch to c, the left-hand edge of the outer part of the cut-off valve; this is equal to the greatest displacement of that valve, and insures that the edge c shall not overrun and contract the port a. From c lay off somewhat more than $\frac{3}{4}$ of an inch, in this case $1\frac{3}{16}$ of an inch, to the left-hand edge of the port b; this gives the necessary length of the bar cd in order that leakage may not occur past the edge c at the maximum displacement toward the right. The clearance, $\frac{3}{8}$ of an inch, is laid off from the right-hand edge of the port b, to determine the edge d of the bar cd. The inner right-hand bar is made as wide as cd. The left-hand half of the main valve and cut-off valve is a counterpart of the right-hand half.

The cut-off valve-spindle takes hold of a lug on one of the bars of the gridiron cut-off valve; and the main valve-spindle passes through a tube or passage cored out through the middle of that valve.

To Set a Double Valve-gear.—Set the main valve to give equal lead; the cut-off by that valve has little influence on the running of the engine, and requires little or no attention. If the cut-off valve is designed to cut off at a definite point when the engine is running under normal conditions, equalize the cut-off by that valve at that point. If the load on the engine and the cut-off are variable within a limited range, the valve should be set to give the least irregularity within that range; it will usually be sufficient to equalize the cut-off for the middle of the range. If the range of cut-off is wide it will often be impossible to get a good action for the entire range, and then it will be advisable to equalize the cut-off for some early point in the stroke of the piston. It has already been pointed out that the Meyer valve may have the cut-off equalized at two points of the stroke by using unequal pitches for the screws on the valve-spindle.

CHAPTER VI.

DROP CUT-OFF VALVE-GEARS.

IN this chapter there will be given descriptions of a few special forms of valve-gears, selected, partly at random, from the large variety of such gears employed by the builders of automatic cut-off stationary engines. All are of the four-valve type of valve-gears, and all give a drop or disengagement cut-off. A description and analysis of these few forms will enable the student to analyze and understand other gears of similar types.

Brown Engine Gear.—Fig. 1, Pl. XXX, gives a section through the head-end valves and valve-chests of the Brown engine; the crank-end valves and gears are a duplication of those for the head end. The admission-valve V is a five-ported gridiron valve on a vertical valve-seat, and the exhaust-valve is a three-ported gridiron valve on a horizontal seat. Both are controlled by valve-gears on the shaft O, which is driven by the engine-shaft through a pair of equal bevel-gears and makes one revolution for each revolution of the engine. It is clear that four such valves might be driven directly by one eccentric on the engine-shaft, or by four eccentrics on the shaft O, and that in such case the four valves would be equivalent to one plain slide-valve, and would be designed by the principles laid down in the first chapter.

The eccentric E, which moves the steam valve-gear, is set to one side of a vertical through e, so that it gives a rapid upward motion to the lever fe. The toe of the lever fe catches under the edge of the latch L, and lifts the valve V through the spindle SV. When the tail of the latch strikes the pin d, the valve is disengaged from the lever fe, and it falls shut; a dash-pot P checks the motion of the valve and prevents jar. The pin d on the arm bd is under the control of the governor through the horizontal shaft b. It is commonly said that the governor on an engine with a detachment cut-off gear has only the light duty of setting the stop (in this case the pin d) that unlatches the gear and releases the valve; the friction of the governor and the attached parts is, or should be, small. Most such gears throw a shock on the governor, tending to disturb it and make it race when the cut-off valve is released; and the governor should be sufficiently powerful to resist the shock. In this gear, when the tail of the latch L strikes the pin d, the shock tends to open the latch and to throw the pin toward the left; both will yield, but the motion of d, and consequently of the governor, is slight.

The exhaust-valve is moved by the cam C, which consists of a groove, in the face of a disk, in which works a roller on the end of the lever trS'. The end S' of the lever is slotted and provided with a block to avoid bending the valve-spindle $V'S'$. The action of this cam is equivalent to that of an eccentric, except that there are periods of rest when the valve is open or shut. Fig. 2 shows two ways of laying out such a cam; it is intended to show general principles only, and would require some modification to fit it to the engine shown by Fig. 1. Let it be supposed that the cam acts directly on the end of a horizontal valve-spindle, such as $V'S'$, Fig. 1, and that its centre is on a prolongation of the path of the valve-spindle. Suppose further that the cam turns toward the left, and that the valve shall begin to open when the line Od is horizontal, and be wide open when Ob is horizontal. To give a uniform motion to the cam, make the curve 1, 3, 7 an arc of an Archimedean spiral; this is done by dividing the angular space bd and the linear space $1', 7$ into the same number of equal parts, and by drawing intersecting arcs and radii, $6'6$, $O6$, $5'5$, $O5$, etc., as shown. The cam from b to c is a circular groove, so that the cam remains at rest till the line Oc comes into coincidence with the path of the valve-spindle. The groove from the line Oc to the line Oa is so designed that it gives a harmonic motion. On the line $1'7$ a semicircle is drawn, and its arc and the angular space cOa are divided into the same number of equal parts; arcs and radii are drawn intersecting at 1, 2, 3, etc. Finally, the cam from Oa to Od is a circular arc, giving a period of rest. The second construction, giving harmonic motion, is to be preferred for heavy valves having a rapid motion, in order that they may start and stop easily and quietly; for valves that move slowly and have a large frictional resistance, the first construction may be preferable, but the cam should be modified by rounding the corners at 1 and 7, to avoid a shock at starting and stopping. The positions of the lines Ob, Od, Oa, and Oc may be chosen by the designer so that the time and rate at which the valve opens and shuts may conform to the requirements of his design and to the dictates of his judgment and experience.

The cam in the figure has a symmetry with regard to the axes xx' and yy' that suggests the resemblance of its action to that of an eccentric.

It is neither necessary nor customary to balance valves of the type used on the Brown engine, for they have little pressure on them to produce friction when they are moving, and when they are shut they are at rest. It is usual and advisable to set the exhaust-valve to give compression nearly up to the steam-pressure in the steam-chest, so that the pressure under the steam-valve is nearly equal to the pressure on it at admission. The valve drops shut at cut-off, and after it is at rest the steam-pressure in the cylinder is reduced by expansion. The expansion is carried down to within a few pounds of the back-pressure, so that at release the pressure on the exhaust-valve is not excessive; at compression the pressure in the cylinder rises after the valve is at rest.

A feature common to many detachment cut-off gears can be well shown by reference to Fig. 1, Pl. XXX. Let it be supposed that the eccentric E has no angular advance, and that the valve has no lead; then the valve will open at the beginning of the stroke, and will have its greatest displacement, provided that it is not sooner released, when the piston is at or near half-stroke. If the latch has not then struck the pin d, it will not strike it at all, and the valve will remain connected to the gear, and will close at or near the end of the stroke. It is also evident that giving angular advance to the eccentric and lap to the valve will limit still further the range of cut-off. In this gear, however, the cut-off may be continued beyond half-stroke by giving a negative angular advance to the eccentric and a clearance to the valve. If an engine with a detachment cut-off that is limited to the first half of the stroke is overloaded, there is a liability that a failure to cut-off will occur, in which case the sudden increase of work due to the steam following the piston to the end of the stroke will make the engine run very irregularly.

Corliss Valve-gear.—Of all types of detachment valve-gears, that invented by Corliss has been most widely known and has received the most favor. A modification of this gear designed by Mr. Edwin Reynolds is shown by Plate XXXI, which represents the valve-gear on the intermediate cylinder of the triple-expansion engine in the Engineering Laboratories of the Massachusetts Institute of Technology.

The Corliss type of engine has two steam-chests, S for the supply and W for the exhaust; the latter is separated from the cylinder in order that it may not be chilled by the exhaust steam. This arrangement produces a somewhat rectangular casting containing the steam-chests and the cylinder, at the four corners of which are placed four valves, two of which, V and v, are admission- or steam-valves, and the other two, W

and w, are exhaust-valves. The valve-seats are bored cylindrical and the faces of the valves are turned to fit; the valves bear on half a circle or less, and are so connected to the valve-spindles that they may follow the valve-seats without cramping the valve-spindles. The valve-spindles, which are at right anangles to the axis of the cylinder, project through stuffing-boxes and carry cranks on the ends, by means of which the valves are turned on their seats. The exhaust-valves W and w have their valve-cranks WD and wd connected directly to pins A and a, in a wrist-plate O, which receives a harmonic oscillation from an eccentric on the engine-shaft. The admission-valves take steam on their inner edges as shown at v, and their cranks carry blocks as shown at the crank end. In the figure a section is taken just behind the crank Vh, which is represented by a dotted line only, in order to show the disengagement-latch zTh, which engages the block h and is carried by the bell-crank lever EVT; the lever EVT is connected by the link EB to the wrist-plate. The latch is opened when the finger Tz strikes the stop x on the ring xr; the ring is placed under the control of the governor through the cut-off rod NM and the double-armed bell-crank lever Mlm, to which the governor-rod is attached at l. The linkage made up of the valve-crank, valve-rod, and wrist-plate, for example $Oadw$, is designed to give a slow motion when the valve is closed, and a rapid motion when opening or closing. The figure shows the wrist-plate and valves in mid-position, the eccentric being erect. The exhaust-valve w has its edge on the edge of the port; its crank moves through the angle dwd_2, while the wrist-pin oscillates through the angle aOa_2, but that crank has only the angular motion dwd_1 while the wrist-plate moves through the angle $aOa_1 = aOa_2$. The admission-valves have a similar action as shown at e_1ve_2. If it be supposed that the governor-balls are at their lowest position (at which the disengagement-gear does not act) it will be seen that this gear differs from the plain slide-valve gear in two points: first, it has four valves;

and second, these valves have a more favorable action when opening and closing, on account of the linkages just described.

From a pin at g in the crank Vh, a dash-pot rod represented by the line gi leads to a vacuum dash-pot shown by Fig. 2. This dash-pot has two pistons, p and P; the lower piston fits nicely in a closed cylinder from which air is excluded; the upper piston works in a larger cylinder that is open to the atmosphere through a series of orifices i, i_1, and i_2 and the pipe O. When the valve-crank Vh, Fig. 1, is raised, it lifts the double piston Pp and a partial vacuum is formed under p, while air enters freely, through the orifices i, i_1, i_2, to the annular space under the piston P. When the valve is disengaged, the weight of the dash-pot and the dash-pot rod, aided by the vacuum under the piston p, closes the valve promptly; while the air under the piston P acts as a buffer and prevents a shock. The pipe O is provided with a hollow plug as shown, by aid of which the escape of air through the orifices i, i_1, and i_2 may be regulated.

A large number of detachment-gears have been devised and used by Corliss, and by others who have used this type of valve-gear. The one shown by Fig. 1, Pl. XXXI, was invented by Mr. Reynolds and has the advantage that the latch mechanism is centred on the same axis as the cut-off stop; consequently the finger z always strikes the stop x at the same angle, and the same force is required to disengage the cut-off valve. The block, when disengaged, slides along the plate y. On the return motion the plate y slides over the block till it can snap on to it, under the influence of the spring st.

It is customary to give a small lap to the steam-valves; consequently, as with a plain slide-valve, the eccentric has a small angular advance. With such an arrangement the eccentric-centre will be on the line of dead-points, and the valves will have their greatest displacement when the crank has moved through 90° less the angular advance, and before the piston is at half-stroke. If the detachment-gear has not been released before the valve has received its greatest displacement, the valve will not be disengaged at all, but will remain under the control of the linkage connecting it to the wrist-plate; and cut-off will occur near the end of the stroke, and will be determined by the lap and angular advance as with a plain slide-valve. It is therefore evident that the range of cut-off for the ordinary form of the Corliss gear is from the beginning of the stroke to half-stroke or less. When a longer cut-off is desired, for example, on the low-pressure cylinder of a compound engine, two wrist-plates may be used: one wrist-plate, moved by an eccentric with a small angular advance, has control of the exhaust-valves, and gives release and compression near the ends of the stroke; the other wrist-plate is moved by an eccentric with a negative angular advance, and has control of the steam-valves which have a clearance instead of a lap: with this device the range of cut-off may be extended beyond half-stroke.

Expertness in laying out Corliss valve-gears can be obtained only by experience, with good examples for models. The steam-port may be made from $\frac{1}{15}$ to $\frac{1}{16}$ of the area of the piston, and the exhaust-port may be made $\frac{1}{10}$ to $\frac{1}{11}$ of that area. The exhaust-valve commonly has no lap; the admission-valve has a small lap, $\frac{1}{4}$ of an inch, in Fig. 1, Pl. XXXI. In that figure ve is the mid-position of the steam valve-crank, and $e'c_1'$ is the maximum valve-displacement, equal to the lap plus the port-opening; ve_1 is the extreme position of the valve-crank. The points b and b_1 are found by intersecting the arc b_1b_2 by arcs drawn from e and e_1 with a radius equal to the length of the link BE. The arc bb_2 is made equal to bb_1, and ve_2, the extreme position of the crank when the valve is shut, is found by intersecting the arc e_1e_2 by an arc drawn from b_2 with a radius equal to the length of the link. The linkage $Oadw$ is laid out in a similar way, except that the angle aOa_1 is from necessity equal to bOb. The lengths of the valve-cranks and the radii from the centre of the wrist-plate to the pins a

and b depend partly on the proportions of the engine and partly on the habit and discretion of the designer; the longer they are the less will be the force exerted on the links, ad and be, and on the pins which they connect. The angle veb should be nearly a right angle, so that a rapid opening of the valve may be obtained. The pin C in the wrist-plate receives motion from the eccentric either directly or through a carrier or single-armed rocker that magnifies the throw of the eccentric in about the proportion of $1 : 1\frac{1}{4}$. The chord of the arc $C_1 C_2$, through which C swings, is not longer than the radius OC in order that the angle $C_1 OC$ may not be more than 30°. The linkages $Oadw$ and $Obev$ are to be laid out by trial to give as nearly as may be the desired motion to the valves. It will be noticed that the radius Oa_1 and the link $a_1 d_1$ are in one straight line at the extreme position of the exhaust-valve; and in like manner Ob_2 and $b_2 e_2$ are in one line; should the linkages be carried beyond these positions, a double oscillation would be given to the valve-cranks, which is considered to have a bad appearance. The system of rods and levers connecting the rings xqN and vn with the governor is laid out so that the cut-off may come at the beginning of the stroke when the governor stands at the top of its range of motion, and when the governor is at the bottom of that range the cut-off may come at or after half-stroke, i.e. the valve will not be released. Though it is not always done, it will be advisable for an inexperienced designer to draw the valve-ellipse for the steam- and exhaust-valves. The ellipse, or more properly the oval, will have a form like that shown by Fig. 3. The valve will be found to open rapidly and to nearly its full width early in the stroke of the piston. A line nn' drawn at the distance xn, equal to the lap from the axis xx', will show that the cut-off occurs near the end of the stroke. The valve will be found to be nearly at rest during the greater part of the time when it is closed.

The steam-valve is disengaged when the latch holding it is

released, but cut-off does not occur till the edge of the valve comes to the edge of the port, which is an appreciable time later. In Fig. 3, Pl. XXXI, let a represent the point of disengagement; then, under the influence of the dash-pot, the valve falls with an accelerated velocity till it is checked by the air-cushion in the dash-pot. Representing the motion of the piston by abscissæ, and the motion of the valve by ordinates (just as in drawing the ellipse), the action of the valve in closing may be represented by the dotted line abc; the point of cut-off is represented by b at the intersection of this line and the lap-line nn'. The piston-displacements may be readily found from the dimensions of the engine and its speed of rotation, but the forces acting on the valve and its resistances cannot be estimated. The forces are the weight of the dash-pot and attached parts, together with the pressure of the atmosphere on the area of the piston p. The resistances are friction of the valve, of the dash-pot, and of other parts of the mechanism, and the varying pressures under the pistons P and p; the pressure under P is due to the escaping air, and under p to the air beneath it when at its lowest position. Though the line abc cannot be determined by calculation or construction, it may be found experimentally by an apparatus described on page 7 for making an engine draw its own ellipse. The action of the valves of a Corliss engine is commonly investigated by aid of a steam-engine indicator; if the indicator-diagram shows a sharp cut-off, and if the other features are good, the action of the valves is considered to be satisfactory.

Putnam Valve-gear.—The Putnam engine has four double poppet-valves, two for admission and two for exhaust. Plate XXXII shows a section through one of the admission-valves, and its valve-gear. XX' is a casting bolted onto the cylinder-casting. The space SS' is the steam-chest, and the space P leads to the cylinder. The two valves V and V_1 are made of composition and, when closed, rest on composition seats let into the casting. The seat of the valve V is large enough to pass the valve V_1, so that the valves may be readily withdrawn through the hand-hole H. The unbalanced pressure, which must be overcome when the valve is opened, is that on the excess of the area of the upper valve over that of the lower valve. The valve-spindle ab is made of iron to avoid unequal expansion and consequent leakage; for, if the distance between the valves is not exactly the same as the distance between the valve-seats, one or other of the valves will not come properly to its seat.

The valve-spindle ab is stepped into a frame mn, shown in section. The arms gf and gq form a bell-crank lever, one arm of which, under the influence of the spring lk, presses on the frame mn; a pin p' is interposed to reduce friction. The other arm, gf, of the bell-crank lever carries the cam-lever fe, which acts on the frame mn through the interposed sliding-block d and the pin p; this cam-lever is driven by the double cam C. This cam C, and three others, one for the other steam-valve and two for the exhaust-valves, are carried by a shaft which is parallel to the axis of the cylinder and which is driven from the engine-shaft through bevel-gears, so proportioned that the cam-shaft makes one turn for two revolutions of the engine. The figure shows the cam in contact with the cam-lever, and the valves on their seats; the engine is consequently at admission. As the engine moves forward, the cam-shaft turns as shown by the arrow and raises the valves, giving admission of steam, till the cam slides past the corner y of the cam-lever; the valve is then released and falls shut under the influence of the spring kl. The governor-rod takes hold of the pin h at the end of the lever gh. When the speed of the engine increases and the governor rises, the lever hg is thrown down and the cam-lever ef is pushed to the left, so that the cut-off comes earlier. No shock is thrown on the governor when the valve is released, but as the edge of the cam is rounded to avoid cutting the cam-lever, there is a tendency to disturb the governor which the governor must be able to resist. Should the valve fail to close for any reason, the other end of the cam will strike on i and close the valve before the engine makes a return stroke.

The exhaust-cam is shown at A. At each end the cam is cylindrical, so that it holds the exhaust-valve open till near the end of the stroke. The exhaust cam-lever is not placed under the control of the governor, but can be set to give a fixed compression.

Gaskill Valve-gear.—One of the steam-valves for the high-pressure cylinder of the Gaskill horizontal pumping-engine, and part of the valve-gear, are shown on Plate XXXIII. The valve is shown in section by Fig. 2, and the seat is shown in section and half-plan by Figs. 2 and 3. The valve is of the Cornish type and differs from the double poppet-valve only in detail. Like that valve it consists of two valves joined together, the inner valve being small enough to pass through the valve-seat of the outer or upper valve. The unbalanced pressure to be overcome when the valve is opened is that on the difference of areas of the two valves. When open, both valves give admission of steam. The valve-seat S is bolted to the cylinder-casting, and a passage leads directly to the end of the cylinder. The valve is covered by a small cylindrical valve-chest; there are two such chests, one at each end of the cylinder, supplied by a branched steam-pipe.

The valve-gear is shown by Fig. 1, in which E is an eccentric on a shaft parallel to the axis of the cylinder, and driven from the engine-shaft through equal bevel-gears, so that it makes one turn for each revolution of the engine. The eccentric-strap has the cut-off toe a at one end and a lug b at the other. From b the rod bC leads to one end of an equal armed lever, and the valve-spindle d is hung from the other arm; the distance between the rod C and the valve-spindle is several times as far as shown in the figure. The lever il, centred at h, is under the control of the governor through the rod ki.

The eccentric and eccentric-strap with the lever il form a radial-detachment cut-off gear.

Suppose first that the lever hl is thrown so far to the right that the toe a does not touch it; then as the centre of the eccentric describes a circular path around the point O, the point b_0 of the line b_0Ea will move on an arc that sensibly coincides with the axis XX', and the point a_1 will describe an oval $a_1a_2a_3$; the valve meanwhile will remain shut.

On the other hand, if the lever hl is supposed to be so far to the left that the toe a may remain always in contact with its curved end, and if by some means it is prevented from rising from that surface, then the point a will travel on a circular arc nearly coincident with the axis XX', and the point b will describe the oval $b_0b_1b_2$; such an action is of course impossible when the gear is connected up, as the valve is on its seat when the point b is on the axis XX' and consequently b cannot rise above that axis.

With the lever hl in the position shown in the figure, the toe a describes the oval $a_1a_2a_3$ till it comes in contact with the curved surface at the end of the lever; and then a slides along that surface, while the point b describes the arc b_1b_0, and the valve is opened as shown by Fig. 2. When the toe comes to the edge of the surface l, it slips off and the valve is thrown shut by the action of a spring and dash-pot. The toe falls from a to a_1, and the point b_0 returns to b on the axis XX'.

It is evident that this form of valve-gear can give a range of cut-off varying from the beginning to the end of the stroke, and that the release does not throw a shock on the governor. On the other hand, the sudden opening of the steam-valve when the toe a comes in contact with the lever hl throws a shock on the valve-gear that might be troublesome at any but the low speeds at which pumping-engines are commonly run.

INDEX.

Admission, 5
Allen link-motion, 61
Allen valve, 26
Angular advance, 5
Area of steam-pipe and steam-ports, 14
Auxiliary valve-circle, 103

Balanced valves, 27
Bell-crank lever, 11
Brown engine gear, 115

Clearance, 5
Compression, 6
Corliss valve-gear, 118
Crank and connecting-rod, 2
Cut-off, 5
" equalization of, 19
Cut-off valve on back of main valve, 102
" " in separate chest, 97
" " with constant travel, 111
" " with shifting eccentric, 100

Dead-centre, to set engine on, 29
Designing Meyer valve, 106
" link-motions, 67
Diagram, elliptical, 6
" sinusoidal, 7
" Zeuner's, 8
Double-ported valve, 25
Drop cut-off valve-gears, 115
Double valves, 97
" " auxiliary circle, 103

Eccentric and eccentric-rod, 3
Equalization of cut-off, 19, 22
Events of the stroke, 5
Exhaust-space, 4
Expansion and compression, 11

Gaskill valve-gear, 125
Gooch link-motion, 53

Hackworth valve-gear, 93

Joy valve-gear, 93

Lap, inside and outside, 4
Lap and lead, for link-motion, 74
Lead, 5, 16
" of link-motions, 52
Lead-angle, 16
Link-arc, radius of, 42, 54, 67
Link-motion, 39
" " Allen, 61
" " analytical discussion, 43, 54
" " designing, 67
" " for locomotives, 71
" " for marine engines, 68
" " Gooch, 53
" " lap and lead, 74
" " location of reverse-shaft, 76
" " location of rocker, 75
" " " " saddle-pin, 75
" " modifications, 65

Link-motion, open and crossed rods, 41, 54
" " port-opening, 78
" " skeleton model, 72
" " slip, 78
" " Stephenson, 39
" " to set, 80
" " Zeuner's diagram, 51, 58
Link-pins, 65
Loose eccentric, 33

Marshall valve-gear, 91
Meyer valve-gear, 104
" " cut-off at inner edge, 108
" " designing, 106
" " for reversing engines, 109
Model for link-motion, 72
" " " " modifications, 79
" " " " applications, 81

Piston-valve, 24
Port-opening, 78
Ports, 4
" area of, 14
Putnam valve-gear, 123

Radial valve-gears, 88
Radius of link-arc, 42, 49, 54, 67
Reduction of slip of link-motion, 69
Release, 5
Reverse-shaft for link-motion, 66, 76
Rocker, 11
" equalization of cut-off by, 22
" location for link-motion, 75

Saddle-pin, 66
Shaft-governor, 35, 37
Shifting-eccentric, constant lead, 37
" " variable lead, 33
Sinusoidal diagram, 7

Skeleton model for link-motion, 72
Slide-valve, 1, 4
" " problems, 17
" " to lay out, 20
Slip of link-motion, 78
" , reduction of, 69
Steam-pipe, area of, 14
Stephenson link motion, 39
Stroke, events of, 5

To set a double valve, 114
" " link-motion, 80
" " slide-valve, 28
" " an engine on a dead-point, 29
Trick valve, 26

Valve, Allen or Trick, 26
" balanced, 27
" double-ported, 25
" piston, 24
Valve-circle, auxiliary, 103
Valve-ellipse, 6
Valve-gear, Brown engine, 115
" " cam, 116
" " Corliss, 118
" " double, 97
" " drop cut-off, 115
" " Gaskill, 125
" " Hackworth, 93
" " Joy, 93
" " Marshall, 88
" " Meyer, 104
" " Putnam, 123
" " radial, 88
" " Walschaert, 88
Valve-setting, for double valves, 114
" " for link-motions, 80
" " for slide-valve, 28

Walschaert gear, 88

Zeuner's diagram, 8, 51, 58

Plate I.

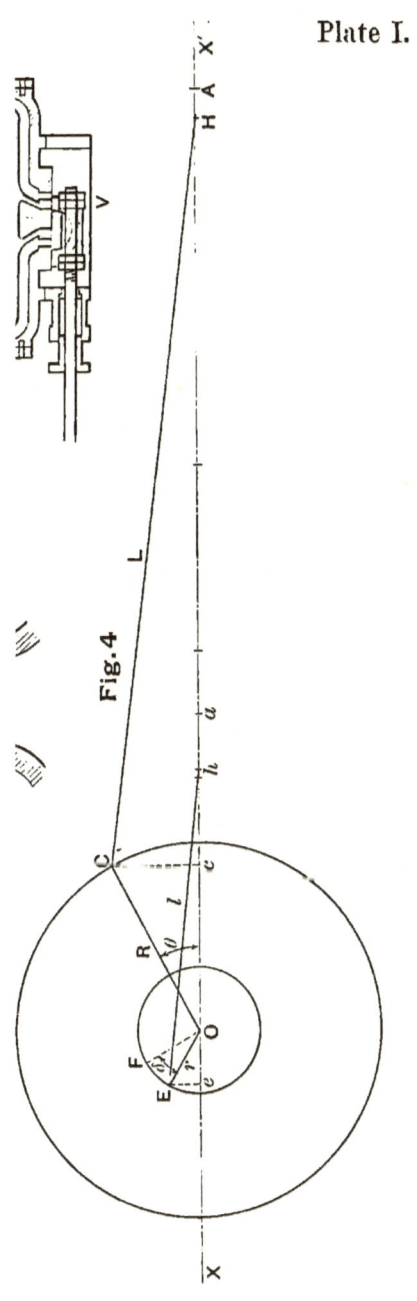

Fig. 4

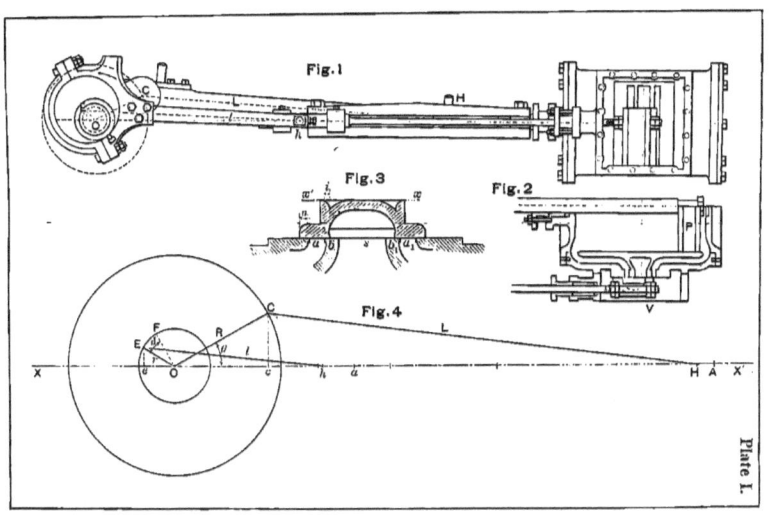

Fig. 2

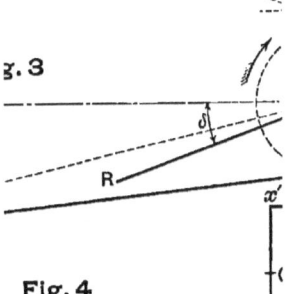

Fig. 3

Fig. 4

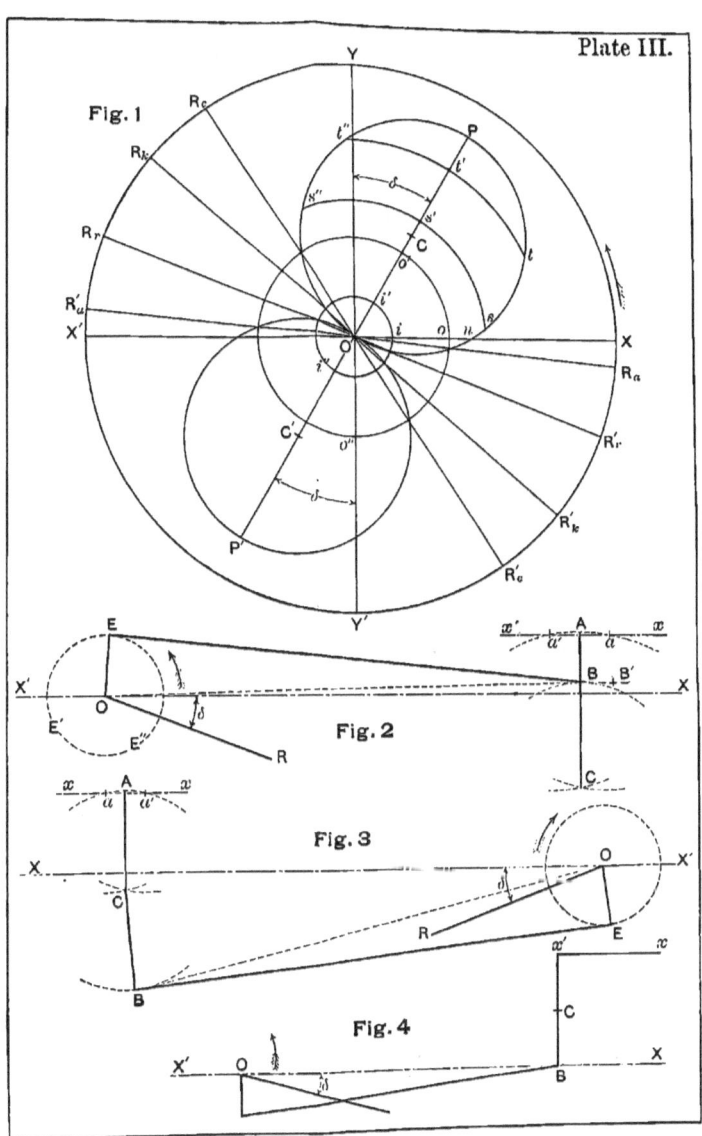

Plate III.

Plate

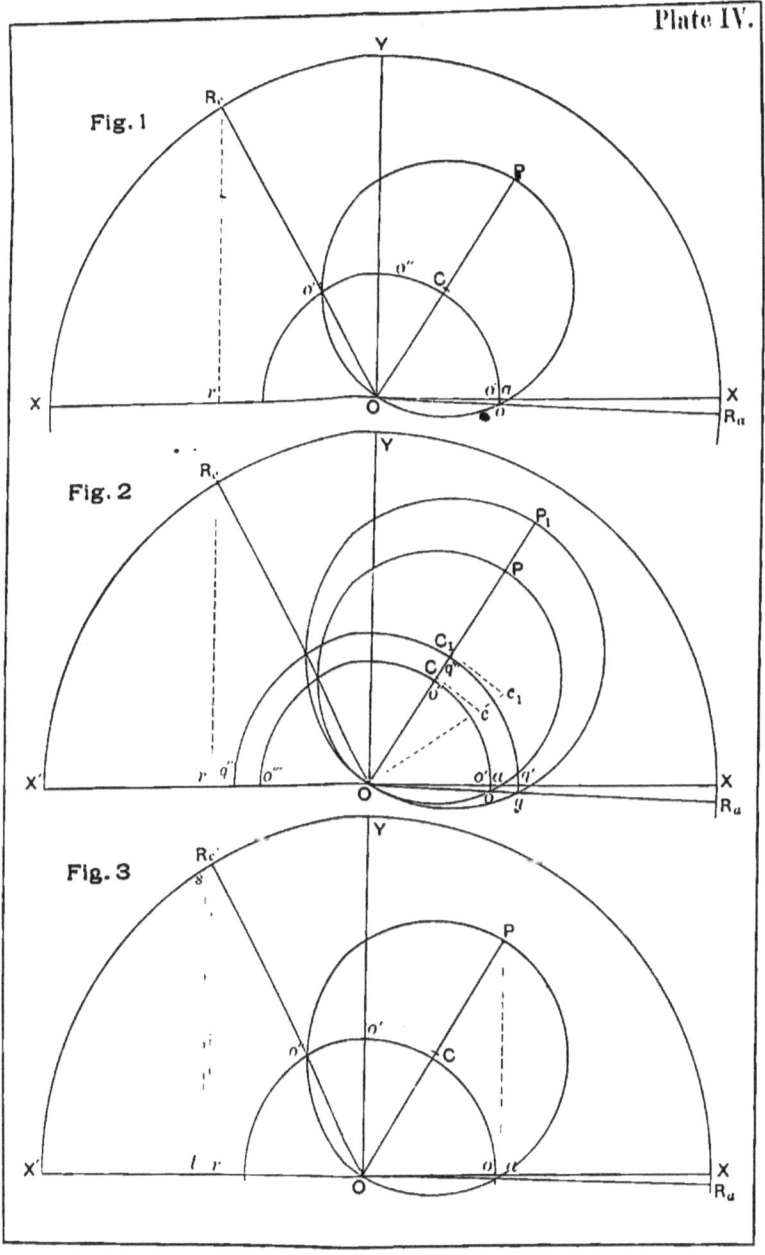

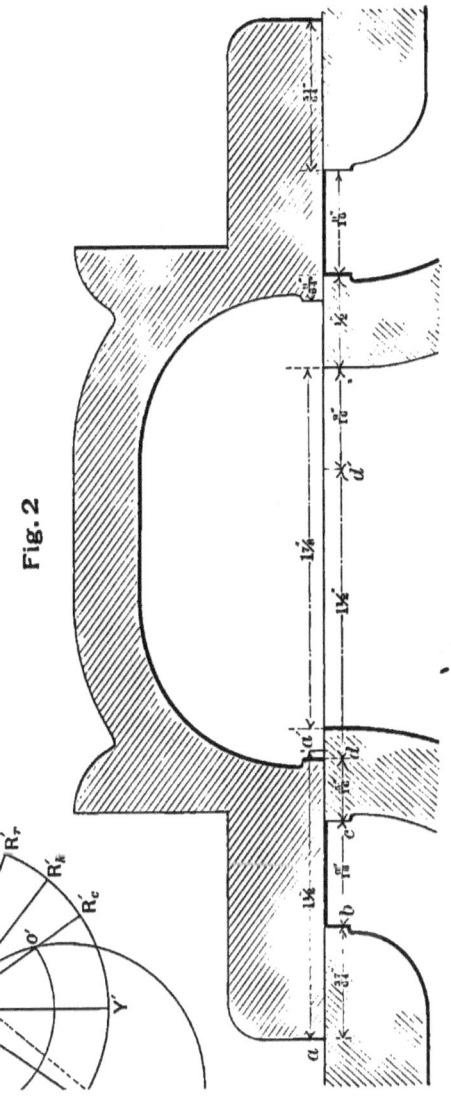

Fig. 2

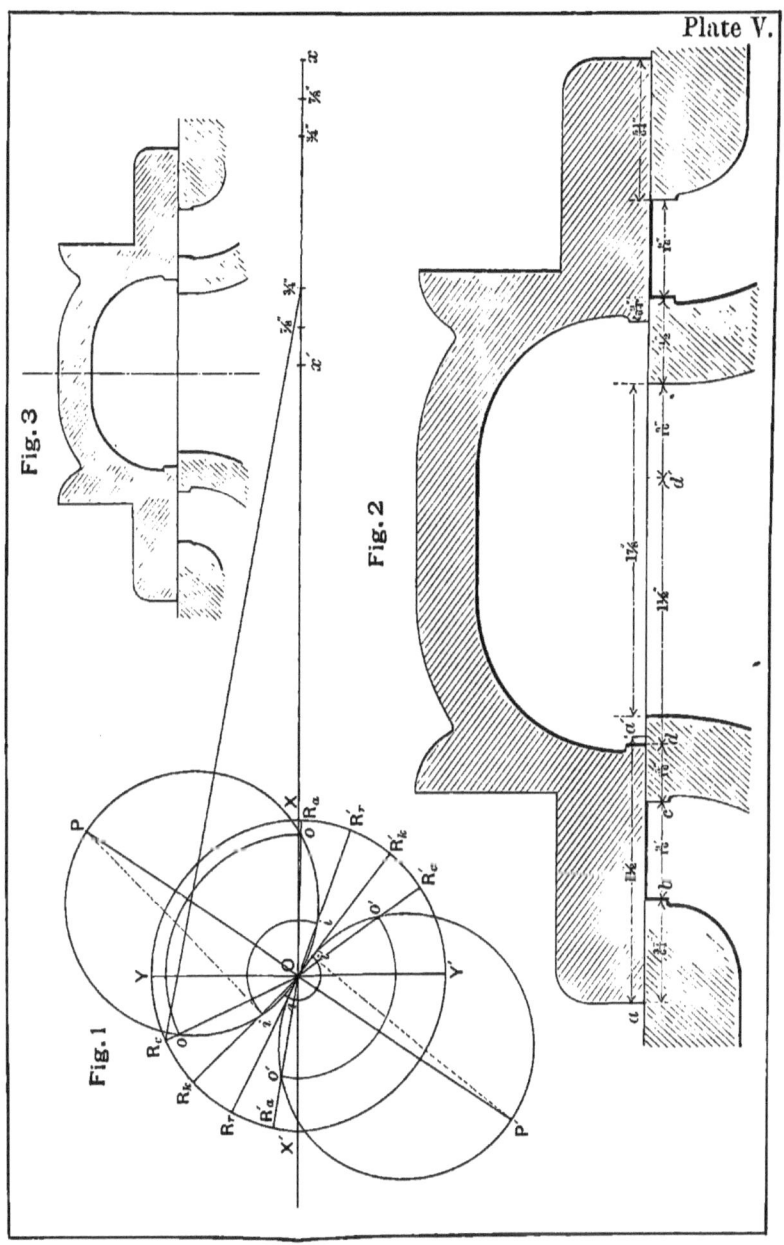

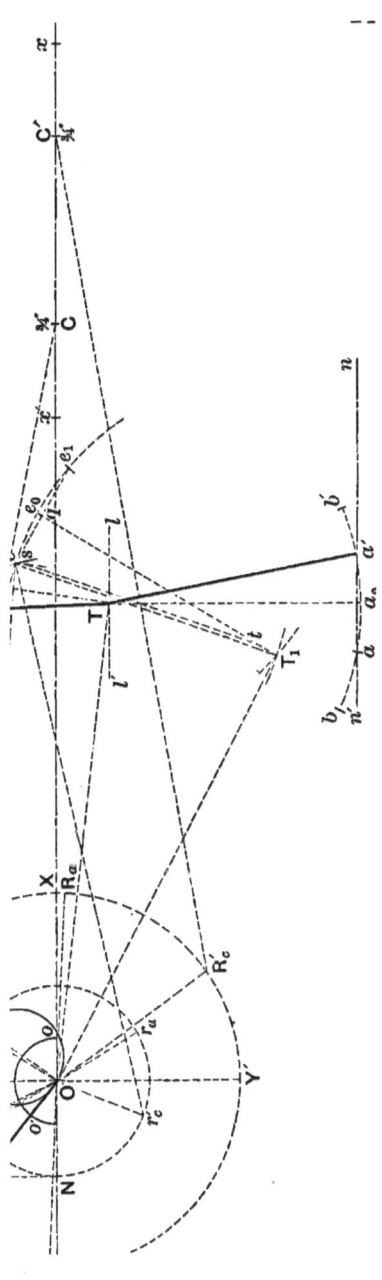

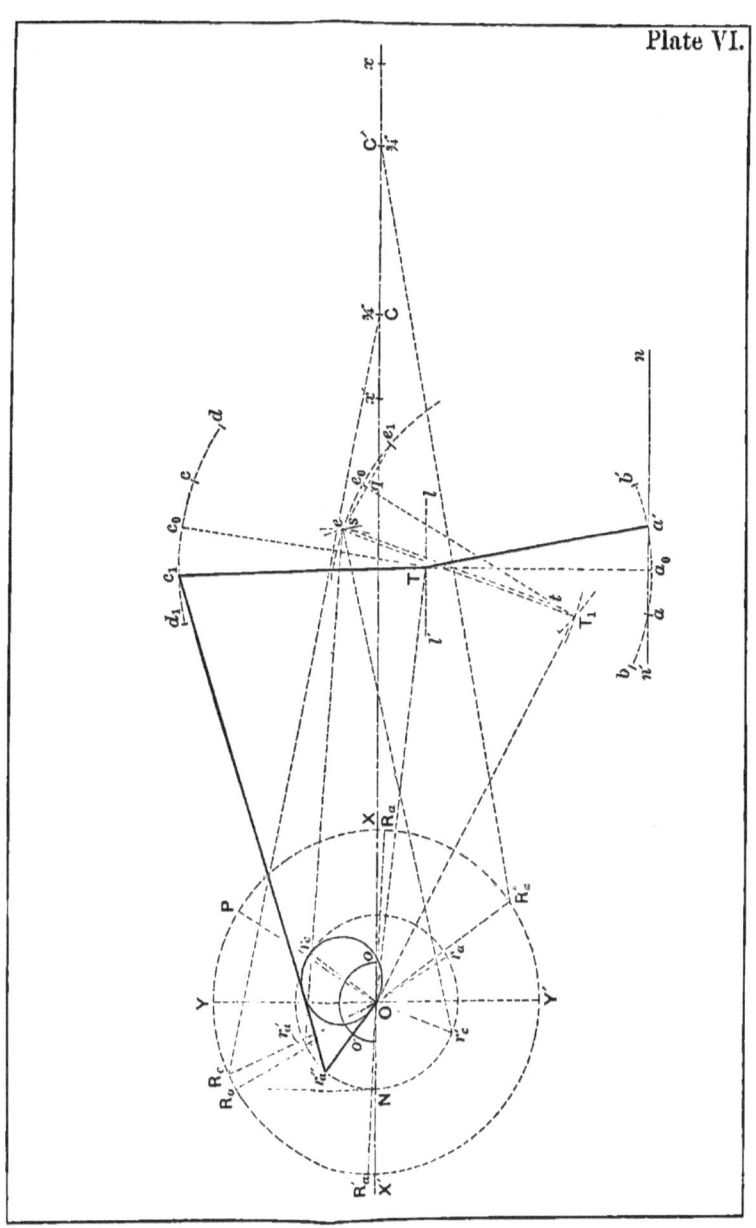

Plate VI.

Plate VII.

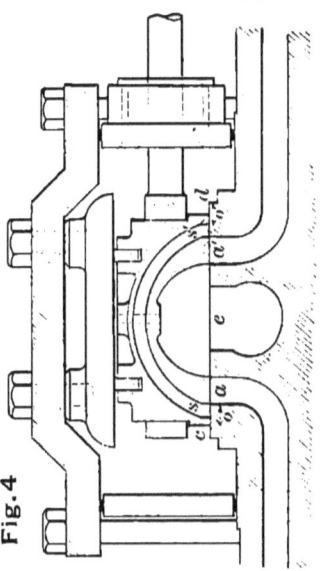

Fig. 4

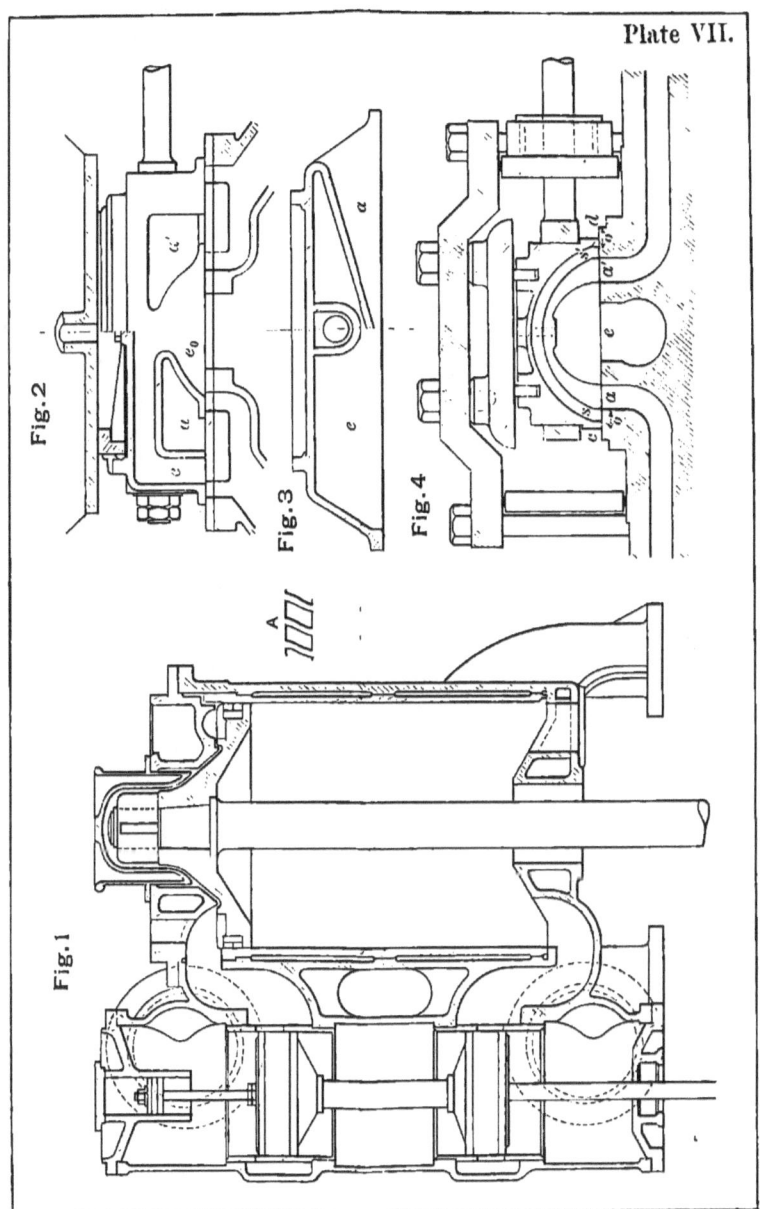

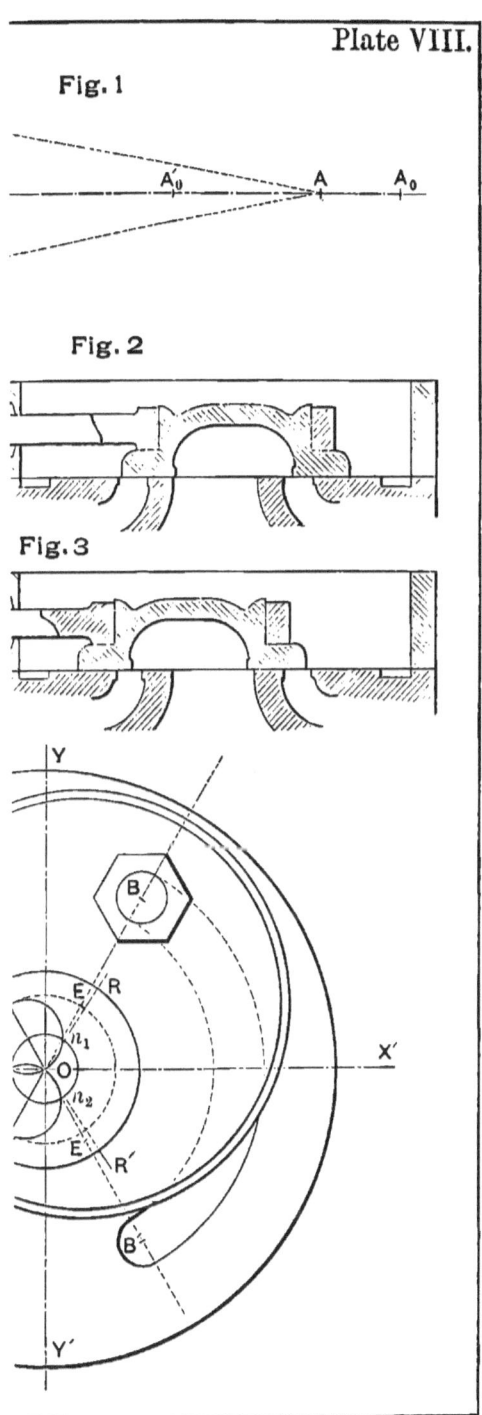

Plate VIII.

Fig. 1

Fig. 2

Fig. 3

Plate VIII.

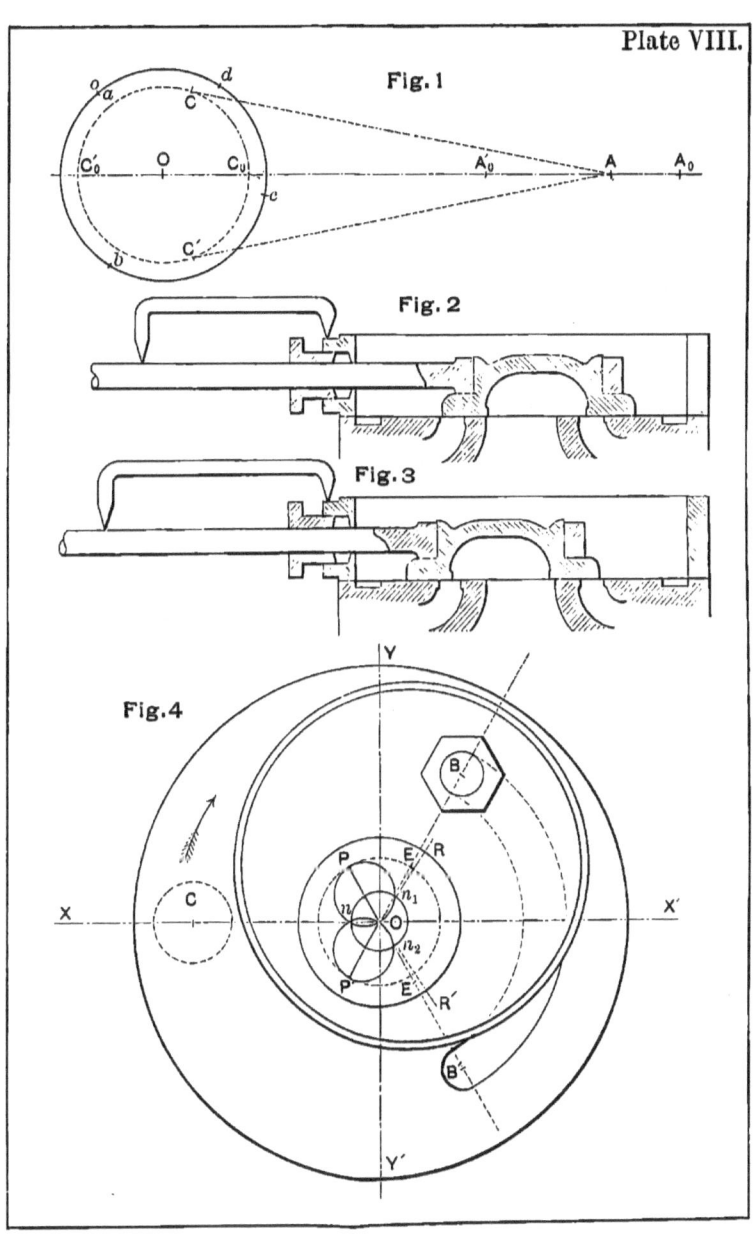

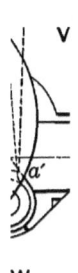

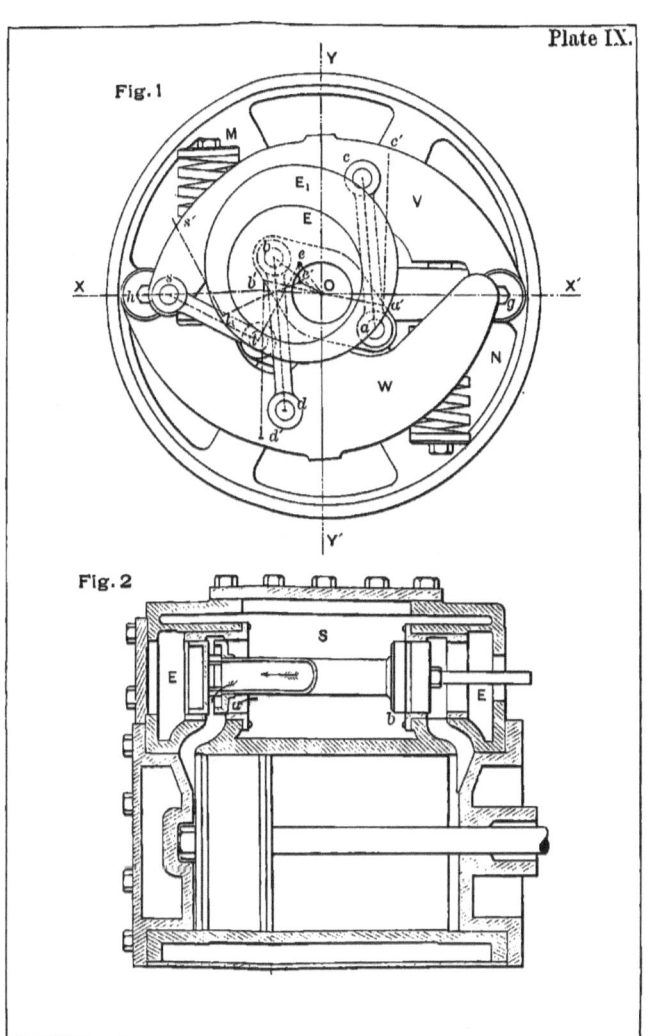

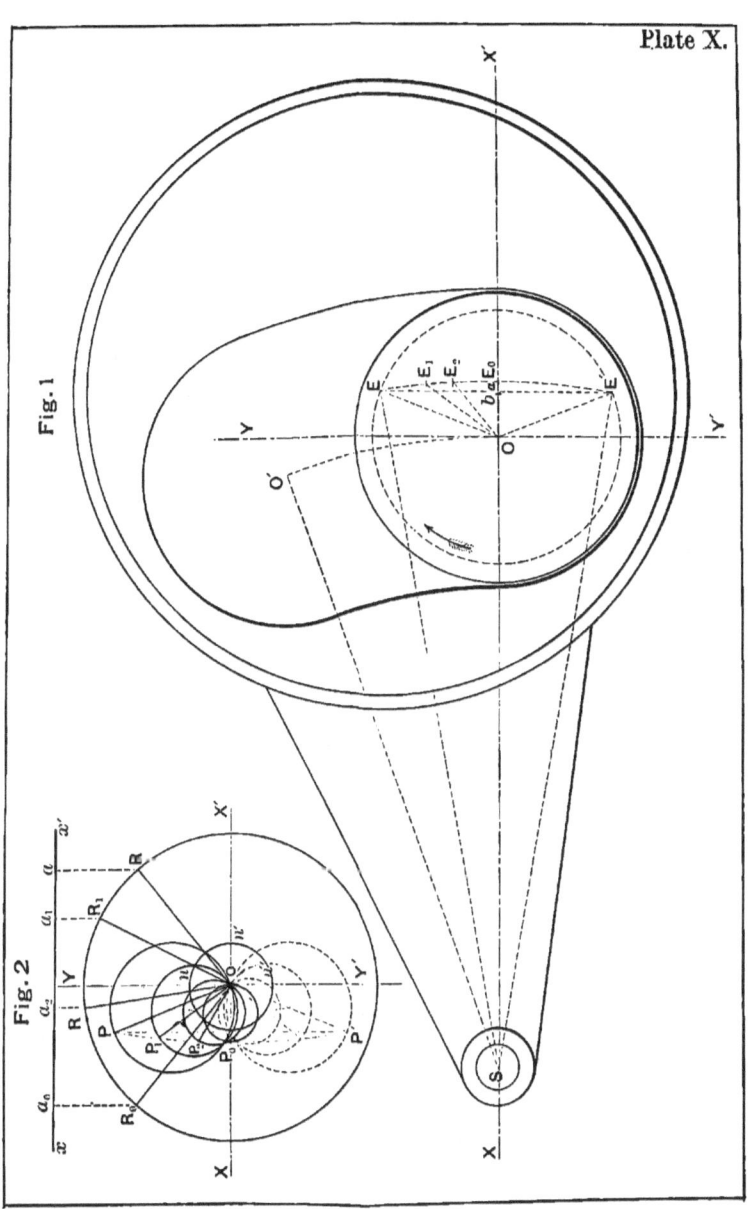

Fig.2

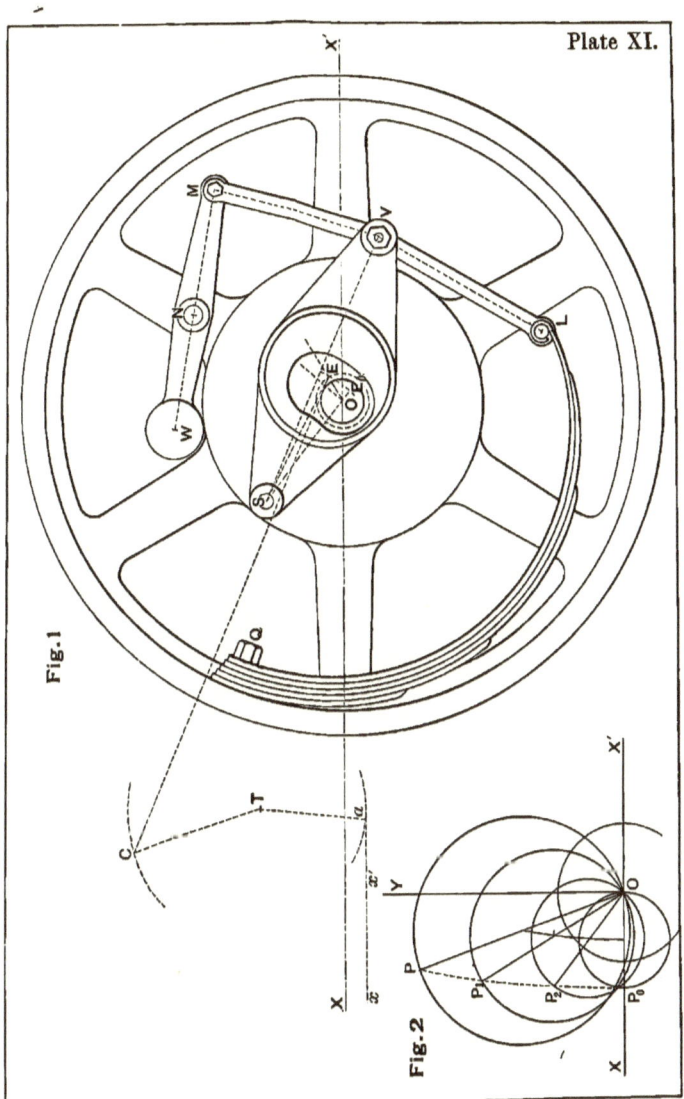

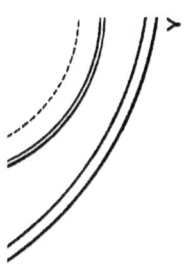

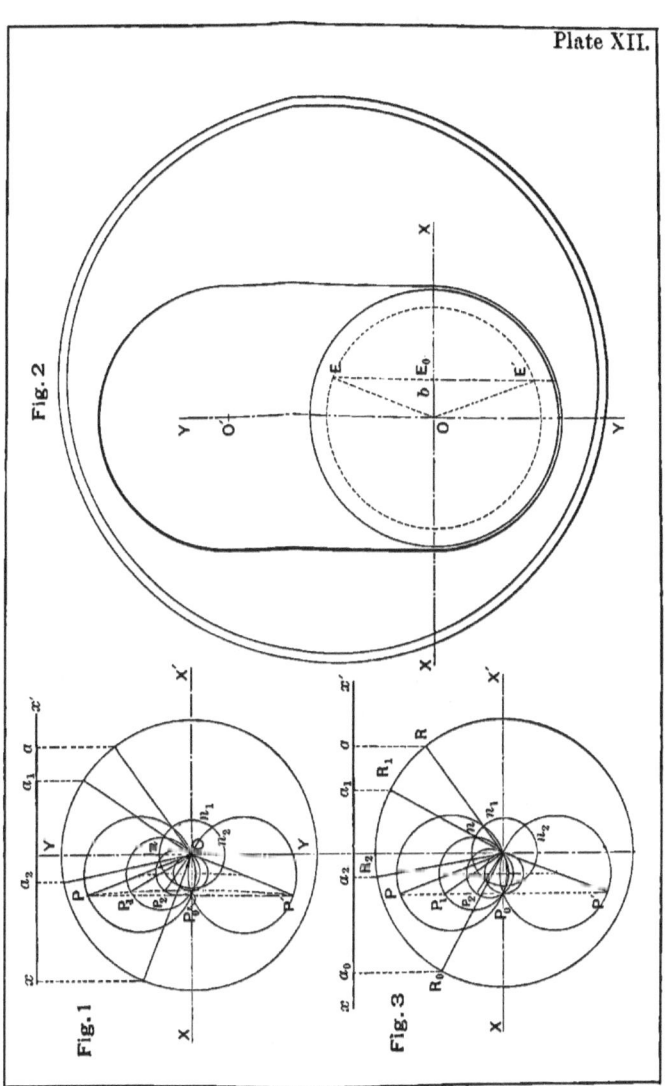

Plate XII.

Fig. 1
Fig. 2
Fig. 3

Plate XIII.

Fig. 3

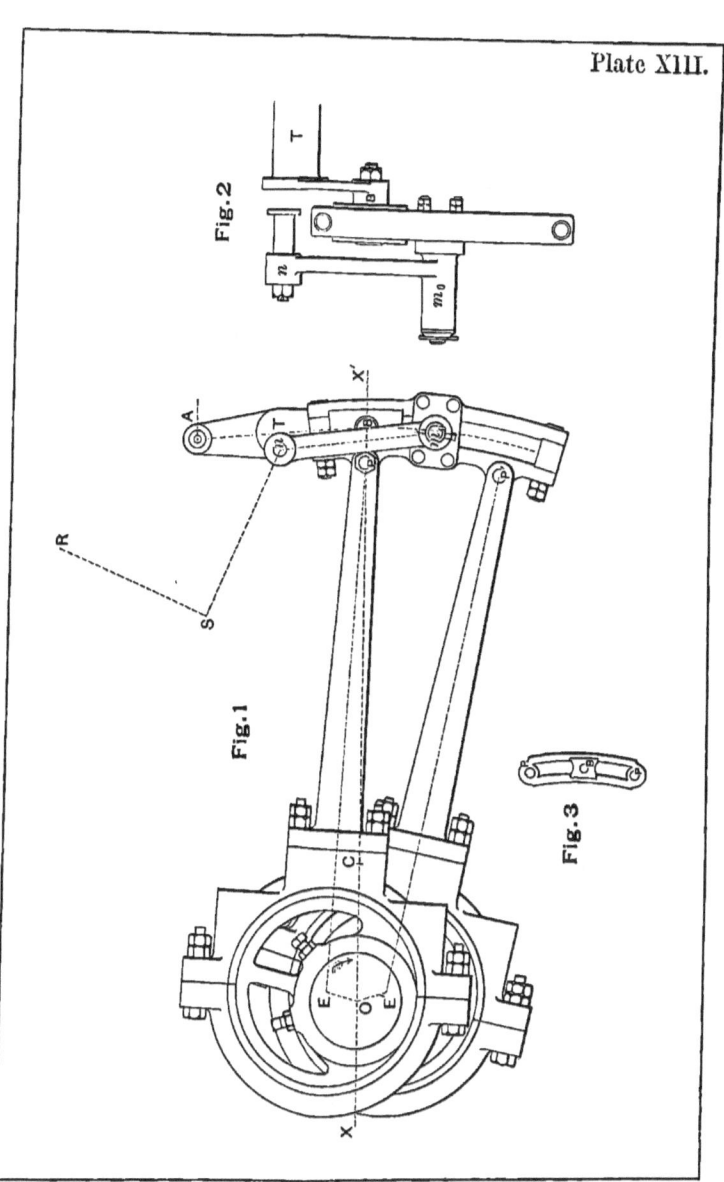

Plate XIII.

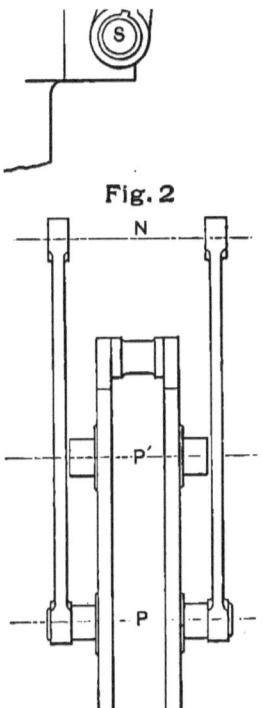

Fig. 2

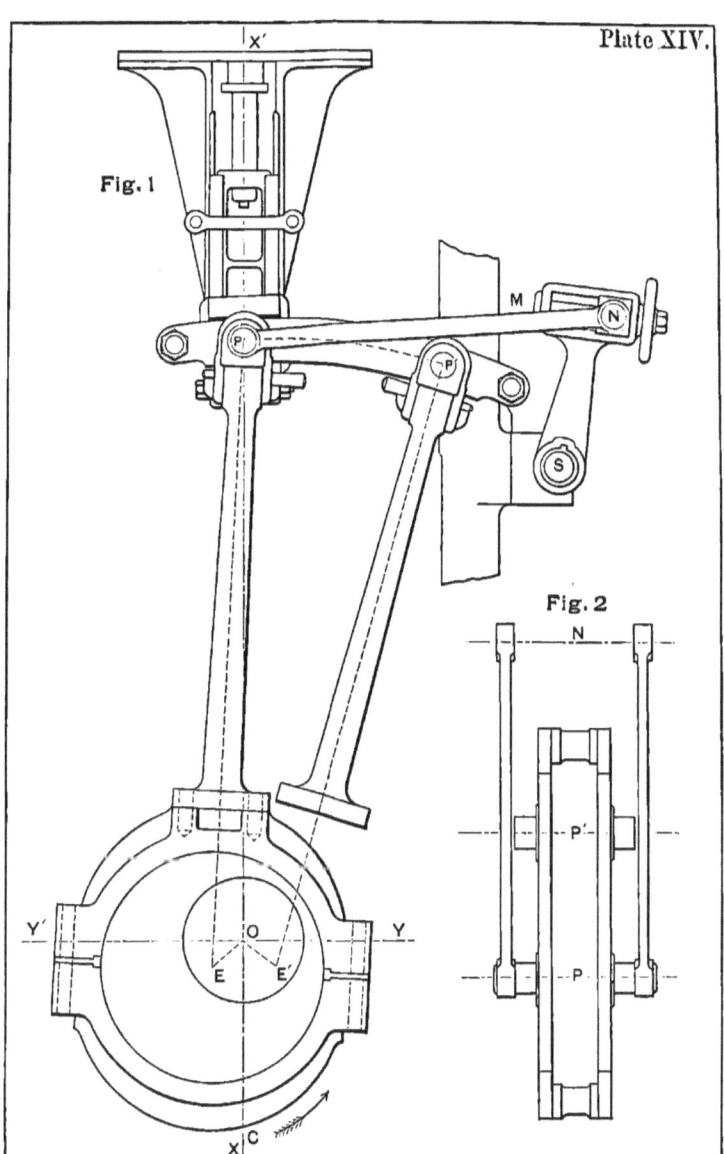

Plate XV.

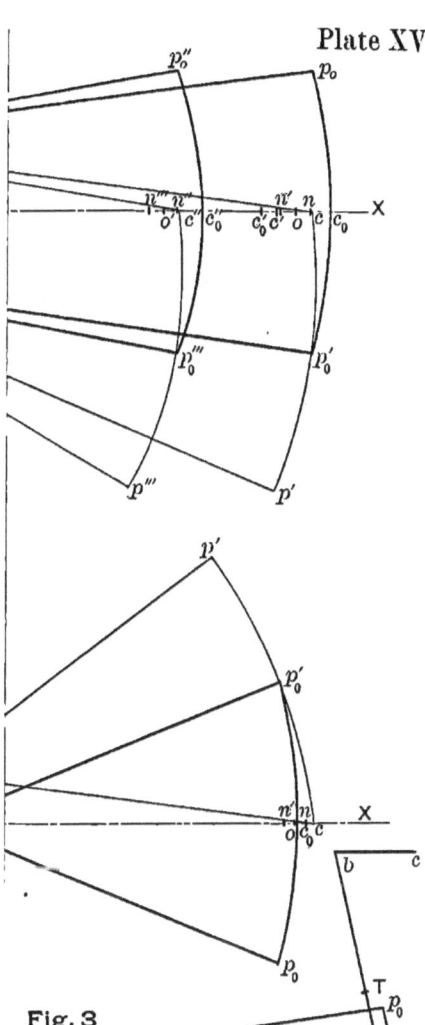

Fig. 3

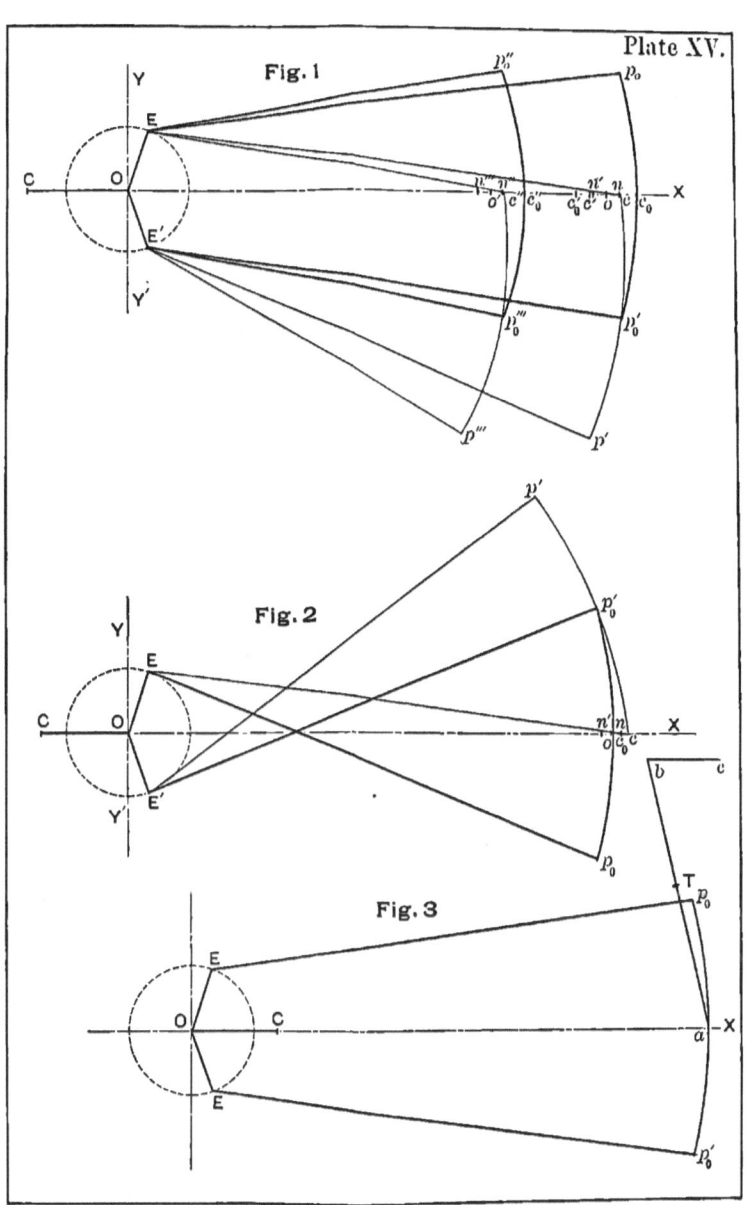

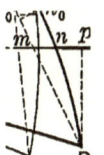

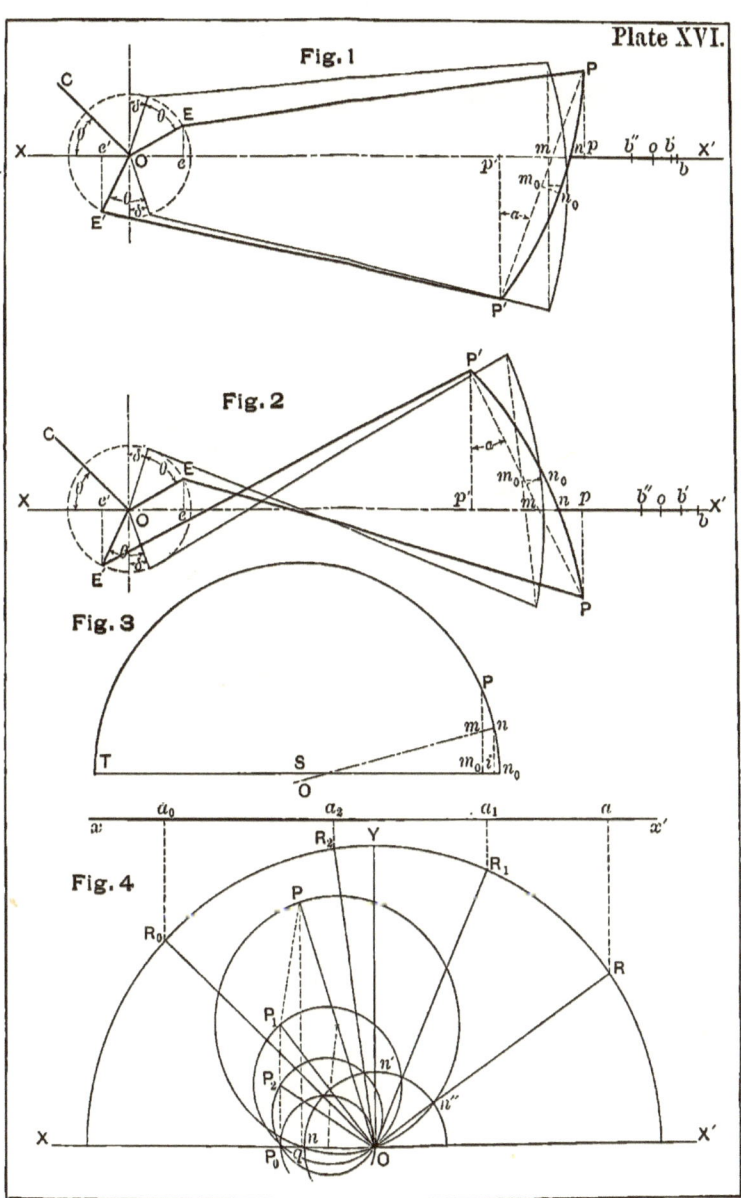

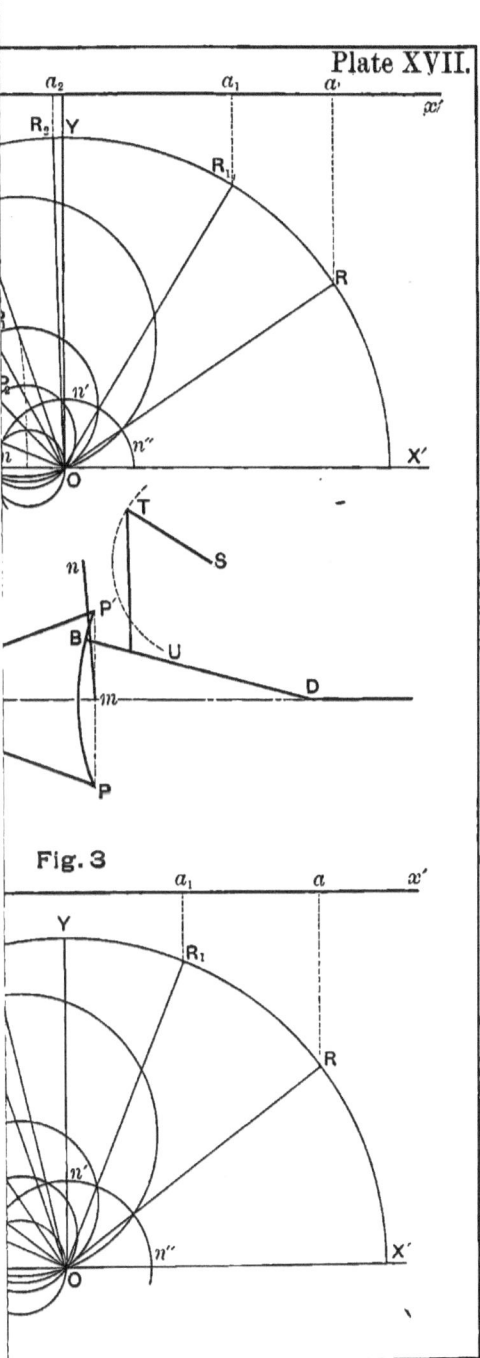

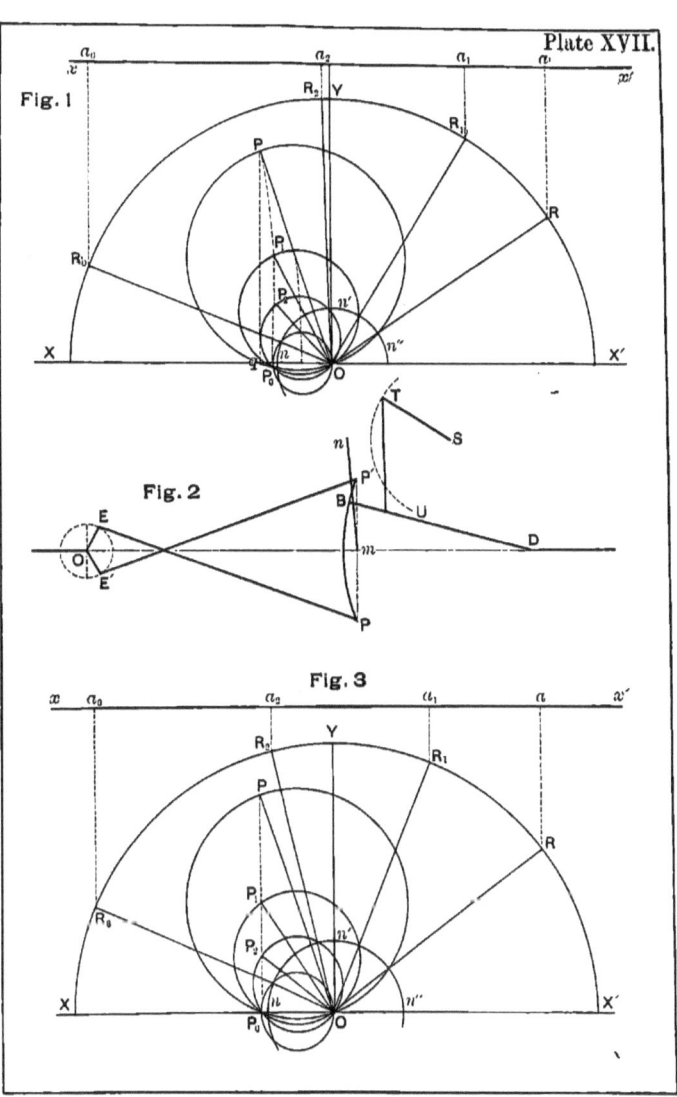

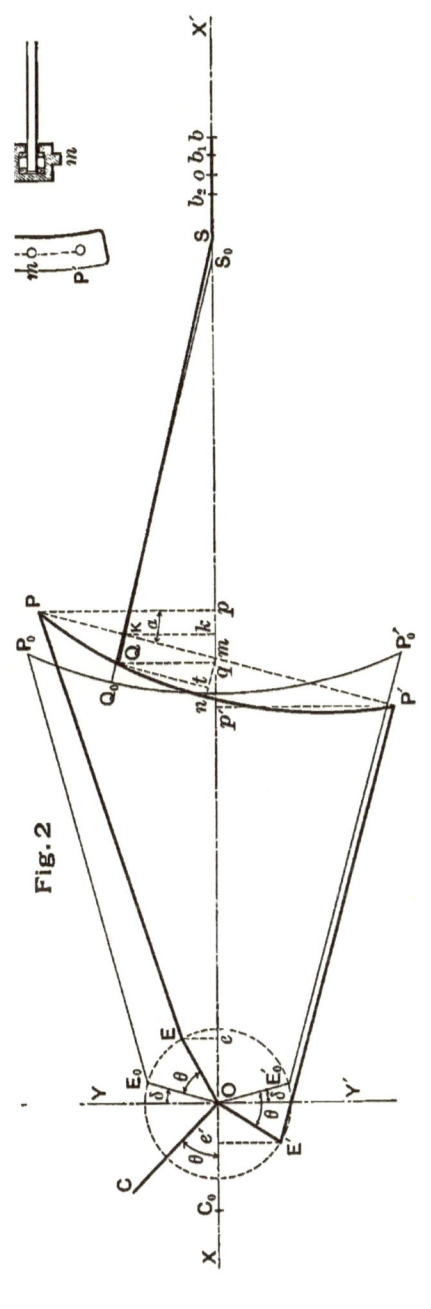

Fig. 2

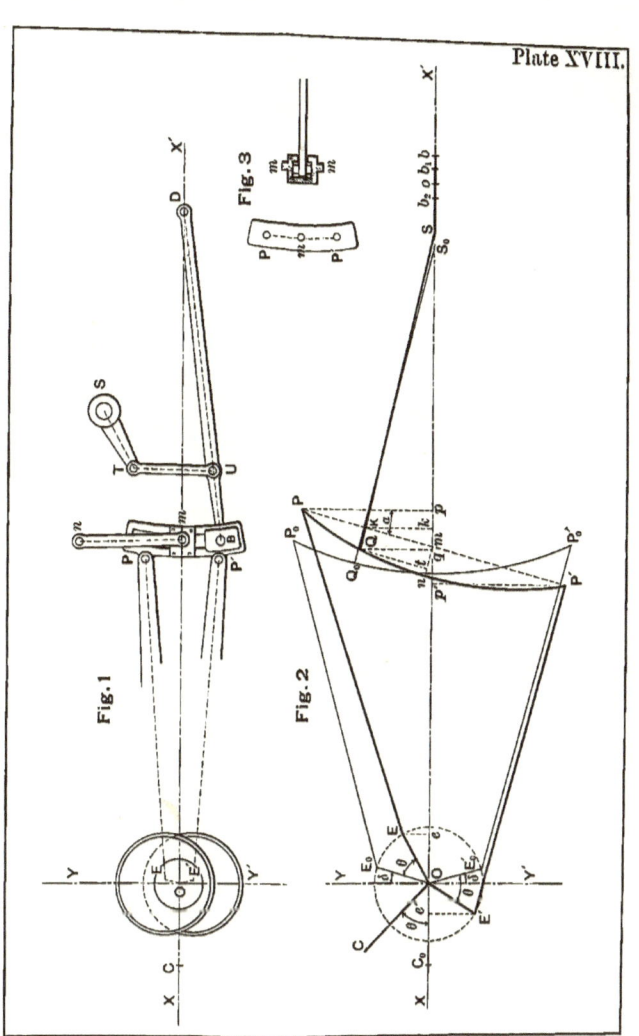

Plate XVIII.

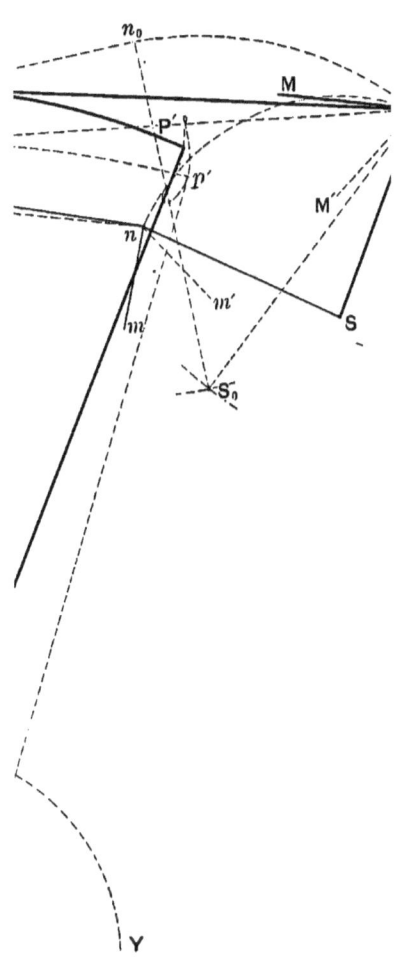

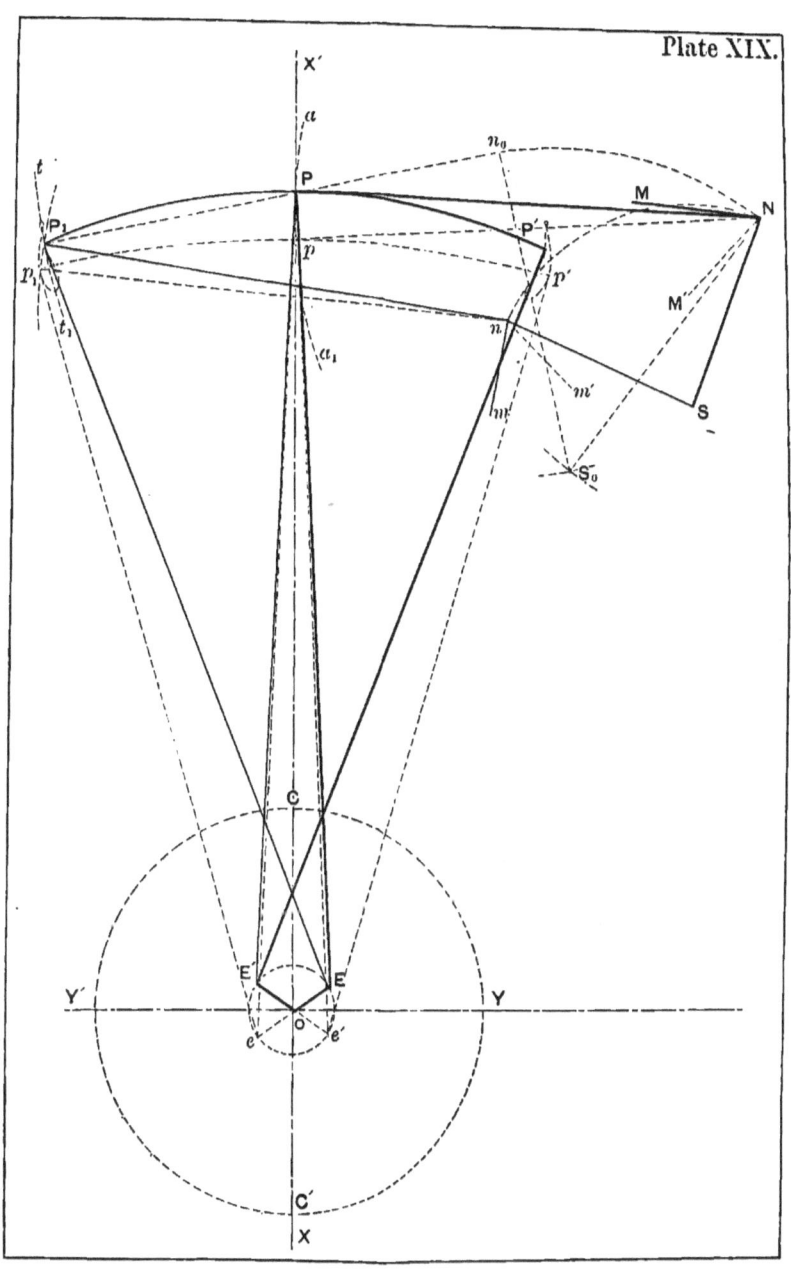

Plate XIX.

Plate XX.

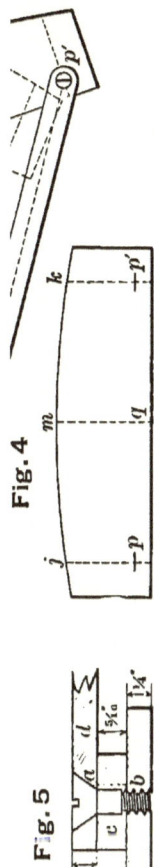

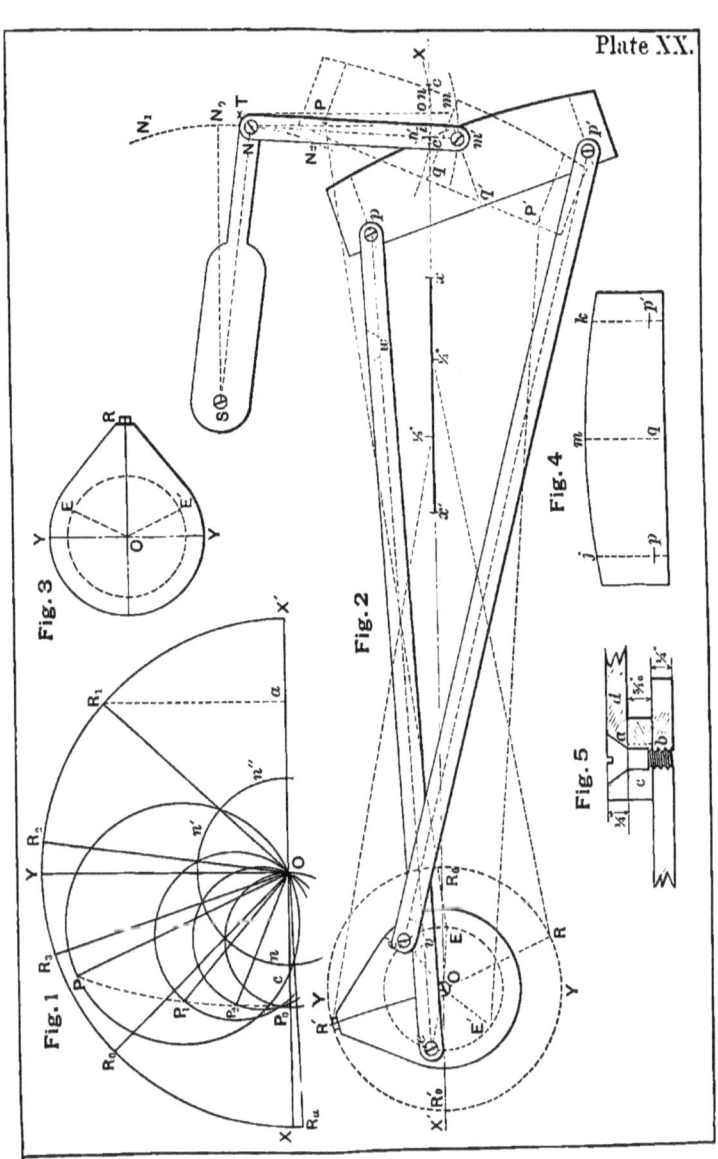

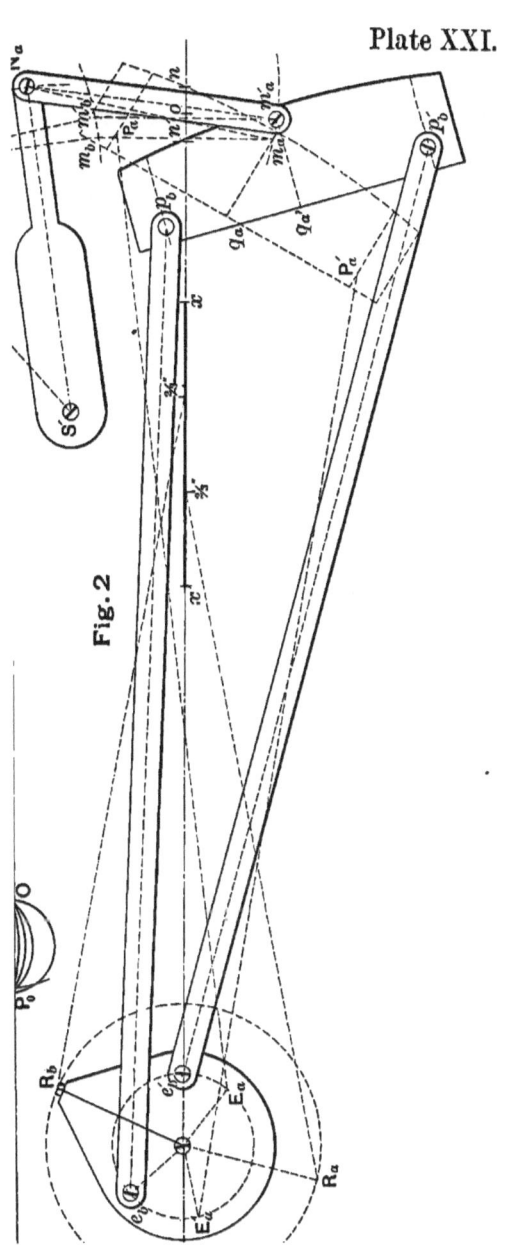

Plate XXI.

Fig. 2

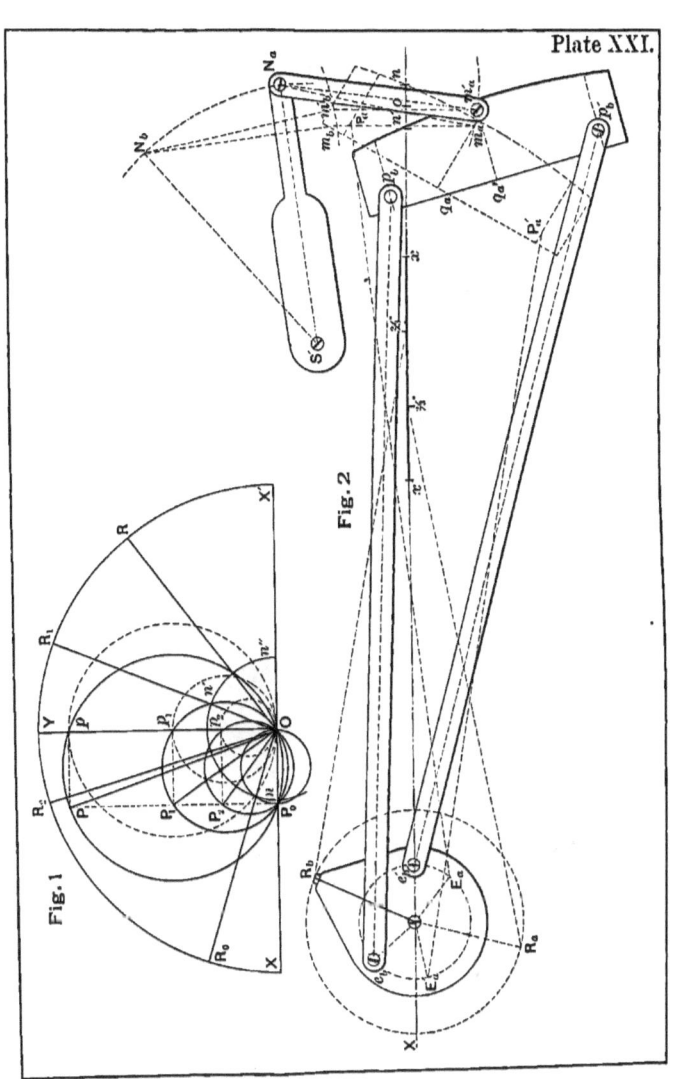

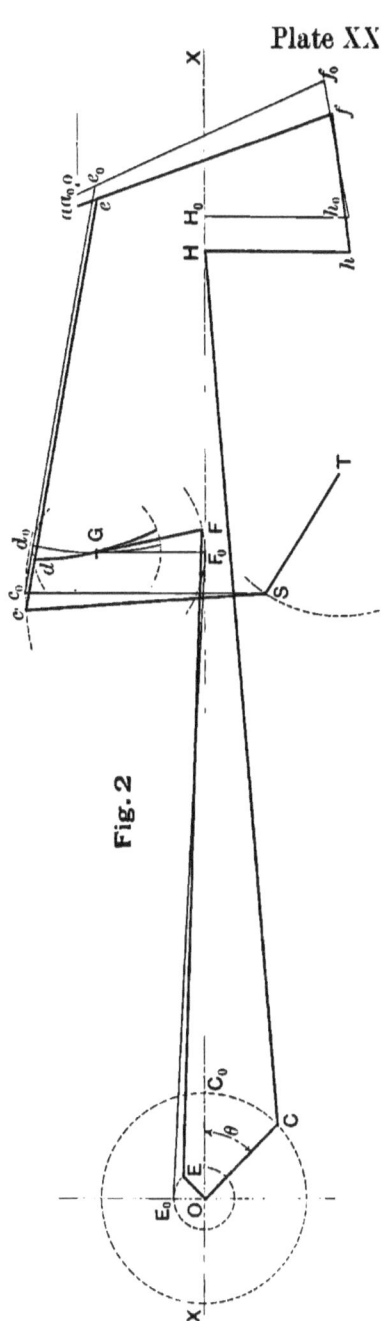

Plate XX

Fig. 2

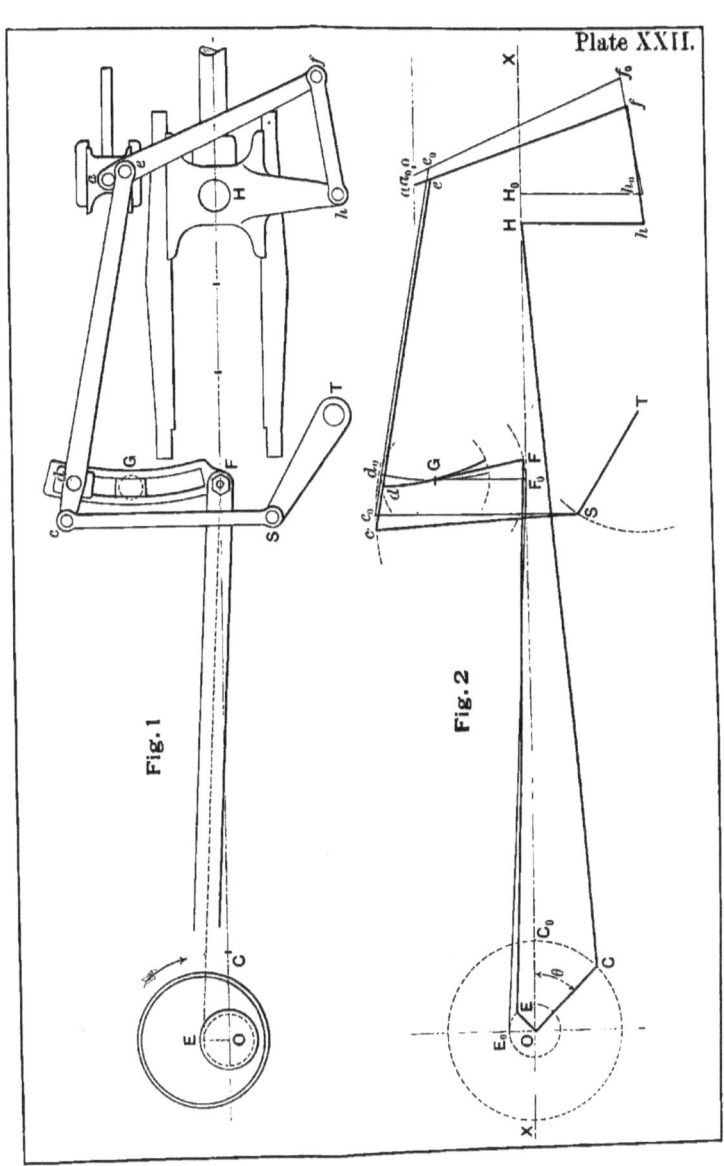

Plate XXIII.

.1

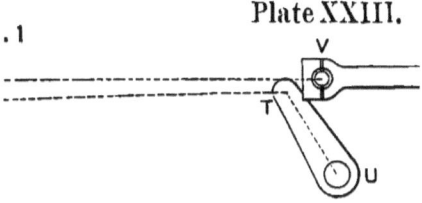

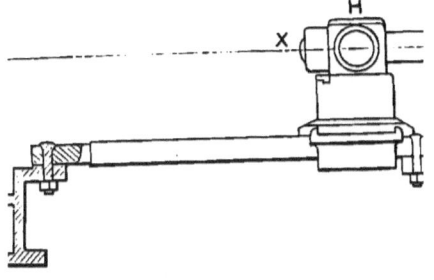

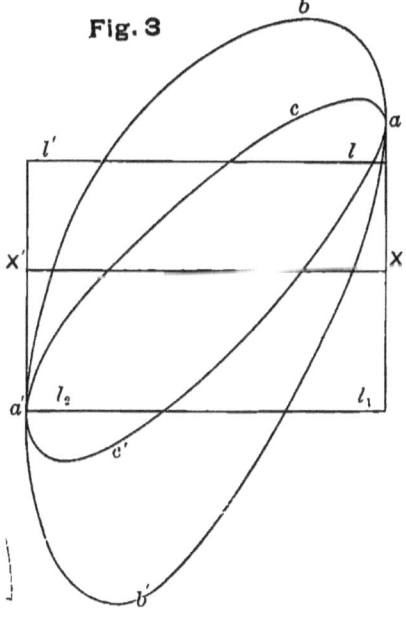

Fig. 3

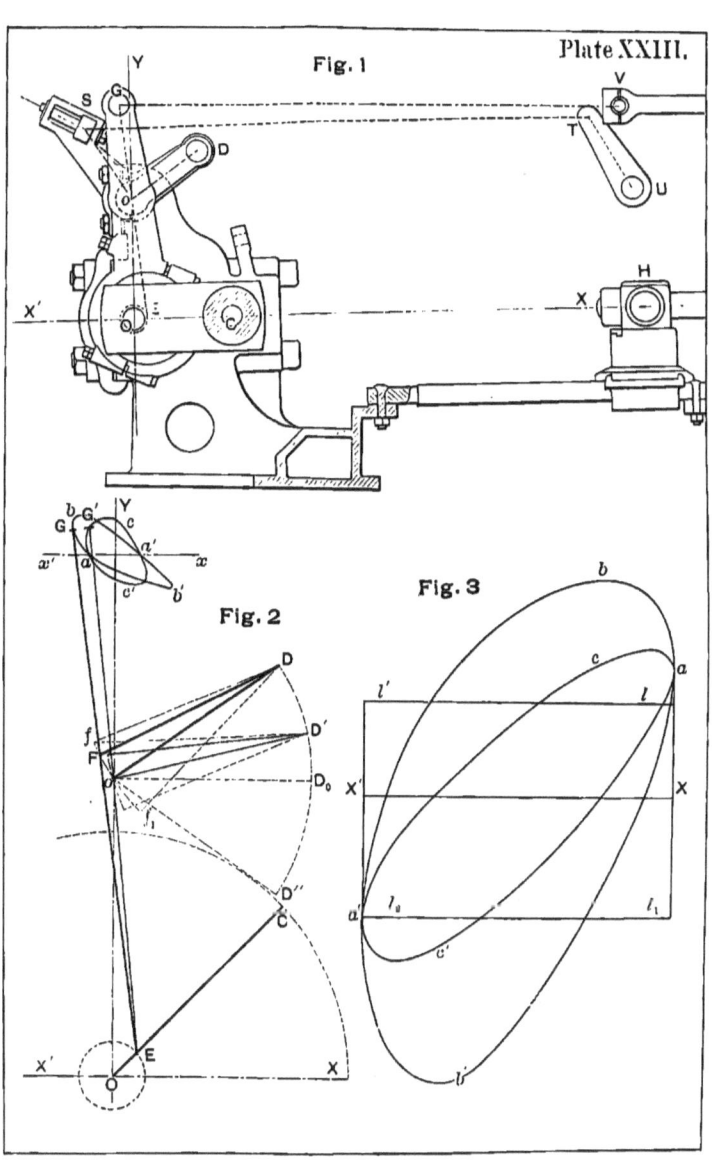

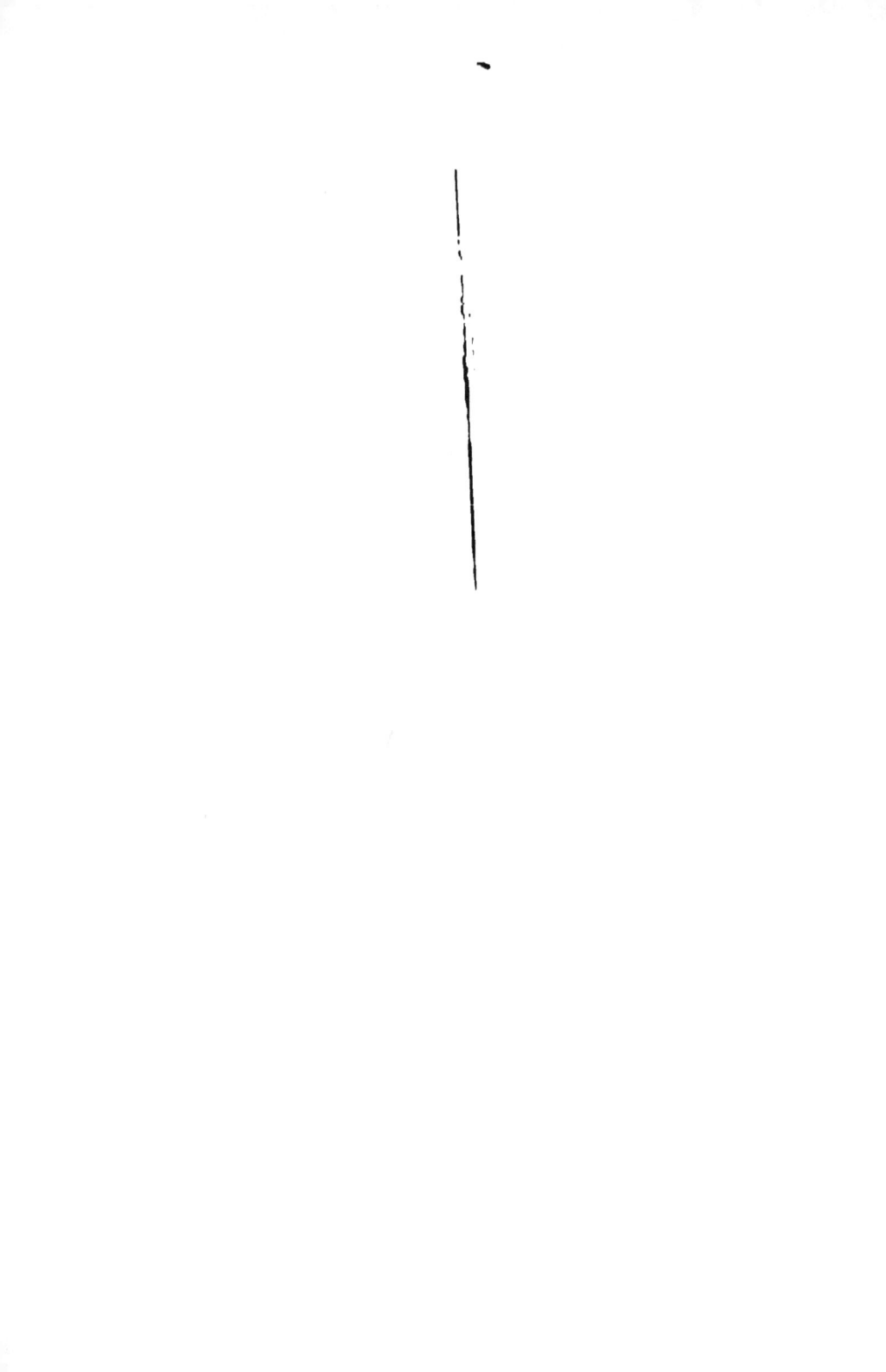

a

$\begin{aligned}&-\\&=\\&-\end{aligned}$

a

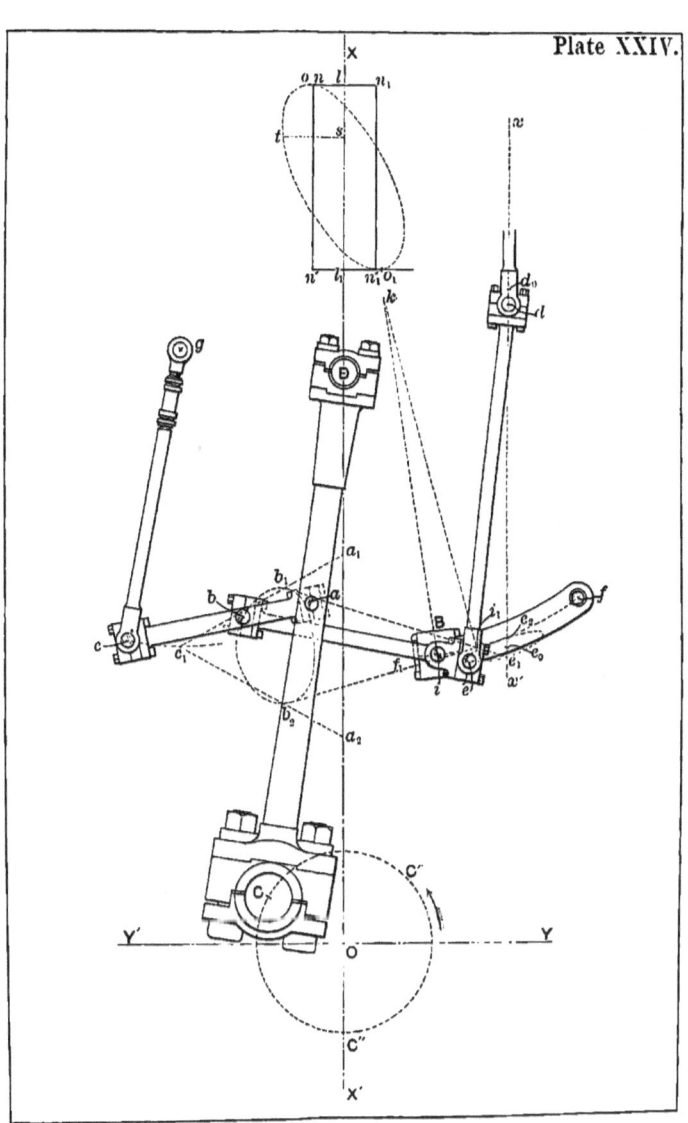

Plate XXIV.

Plate XXV.

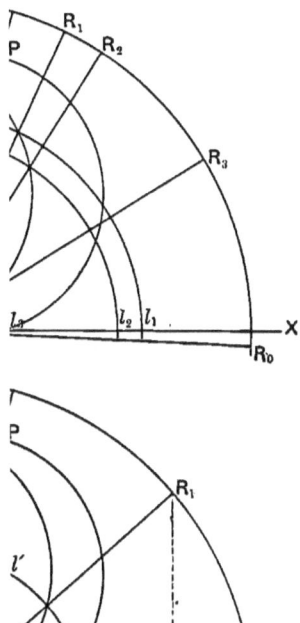

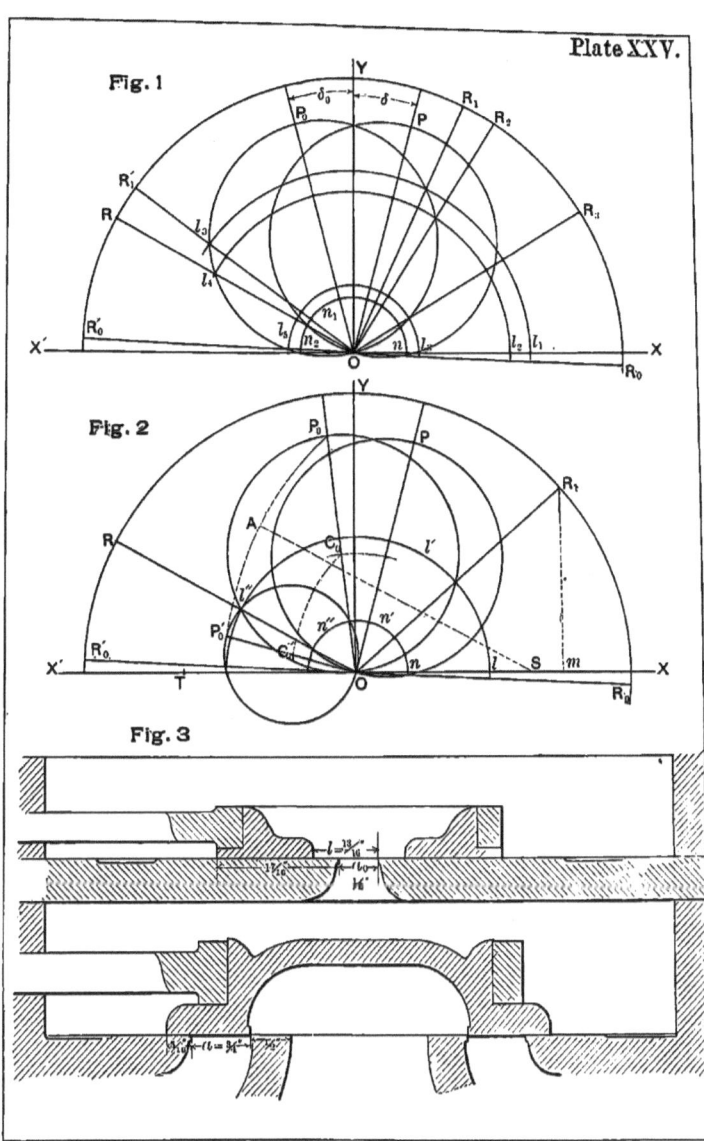

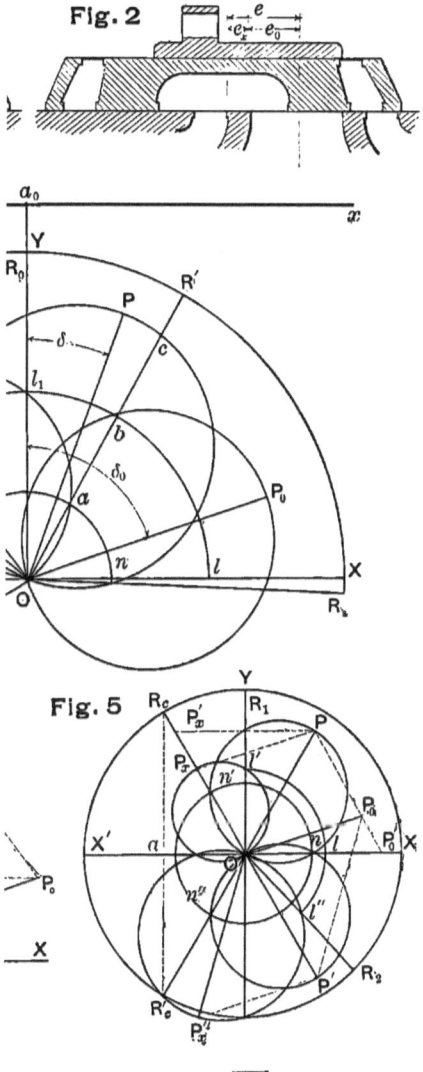

Plate XXVI.

Fig. 2

Fig. 5

Fig. 7

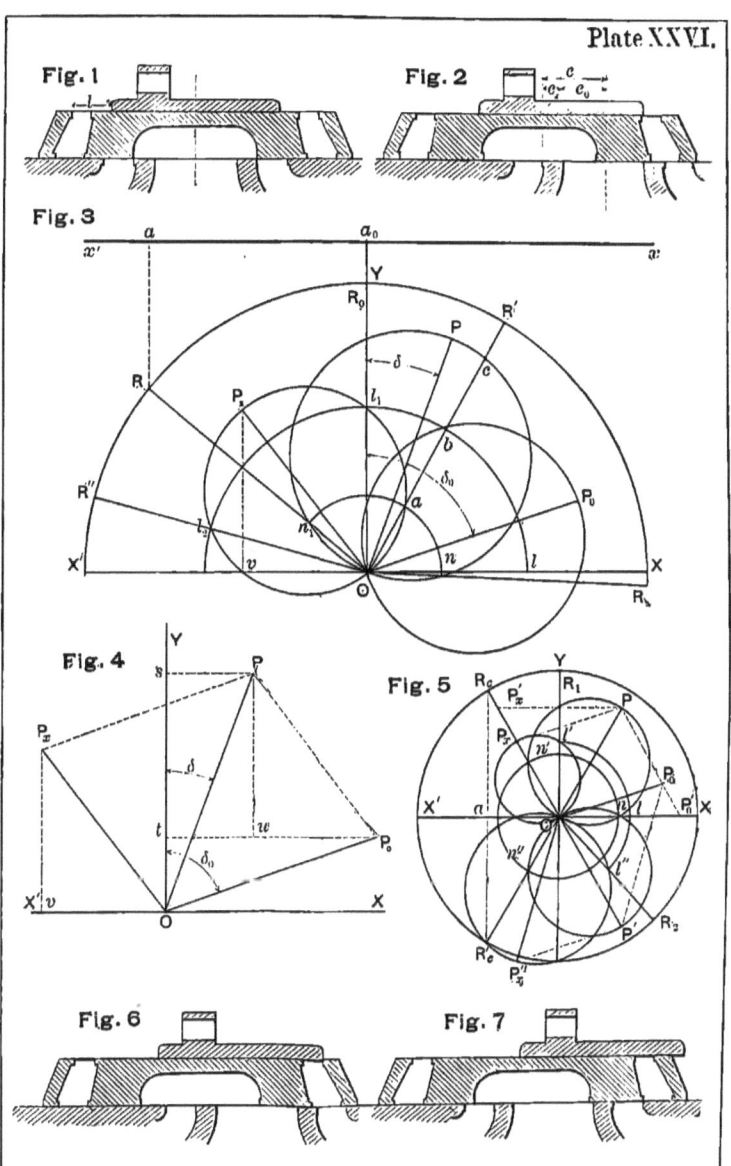

Plate XXVII.

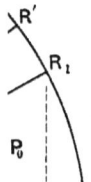

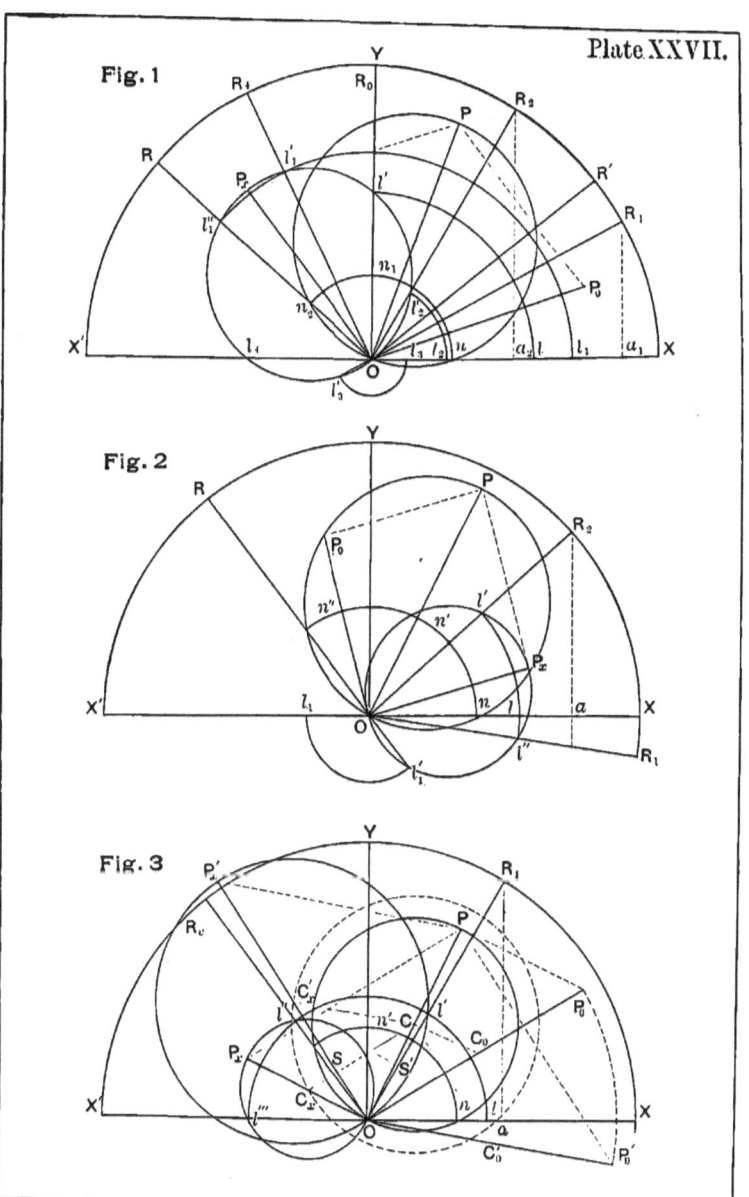

Plate XXVIII.

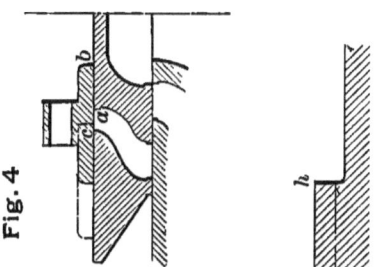

Fig. 4

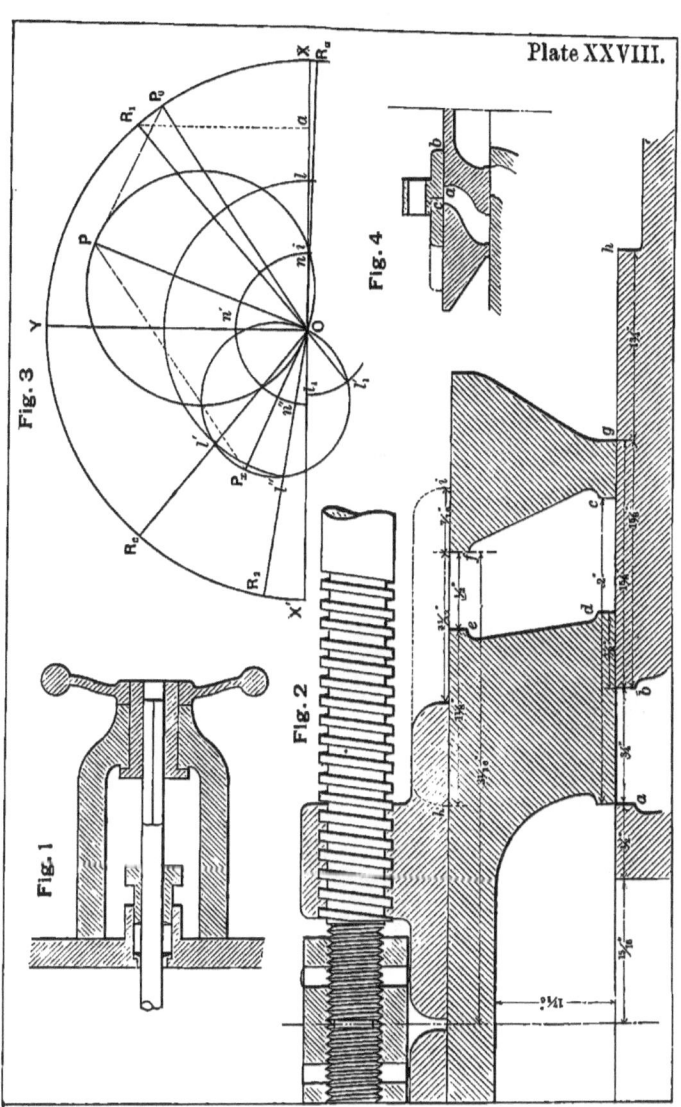

Plate XXVIII.

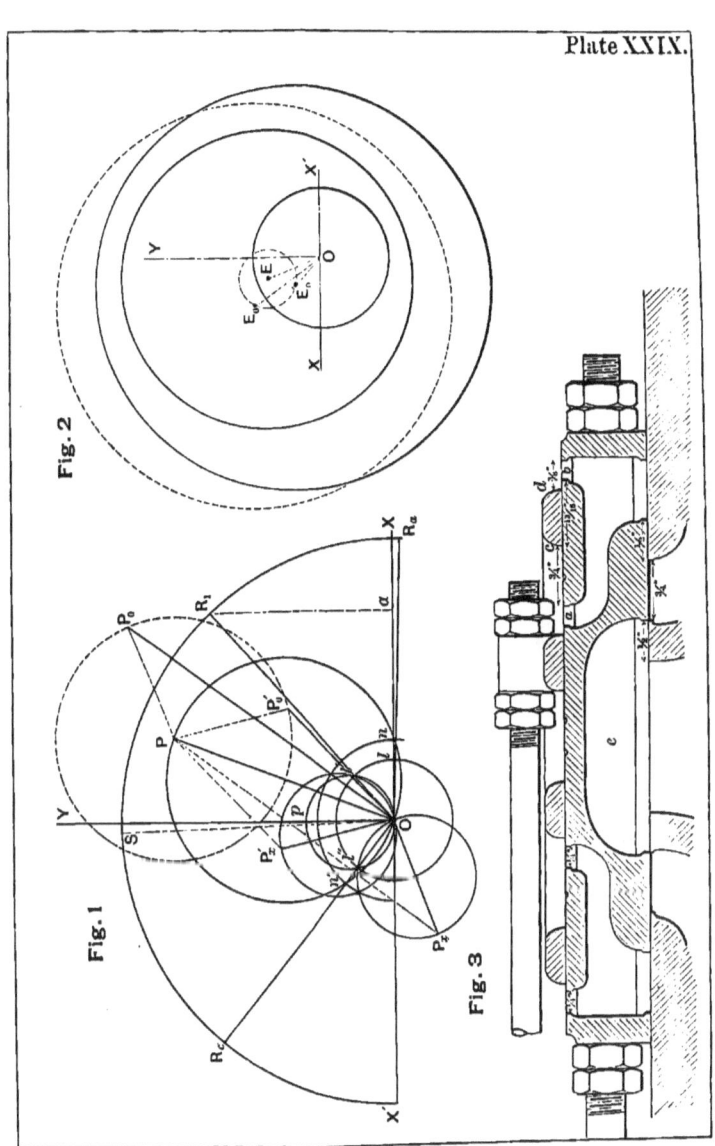

Plate XXX.

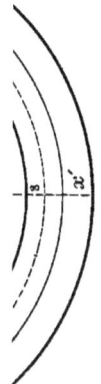

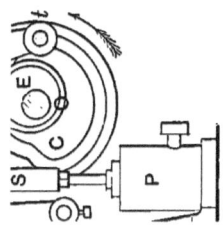

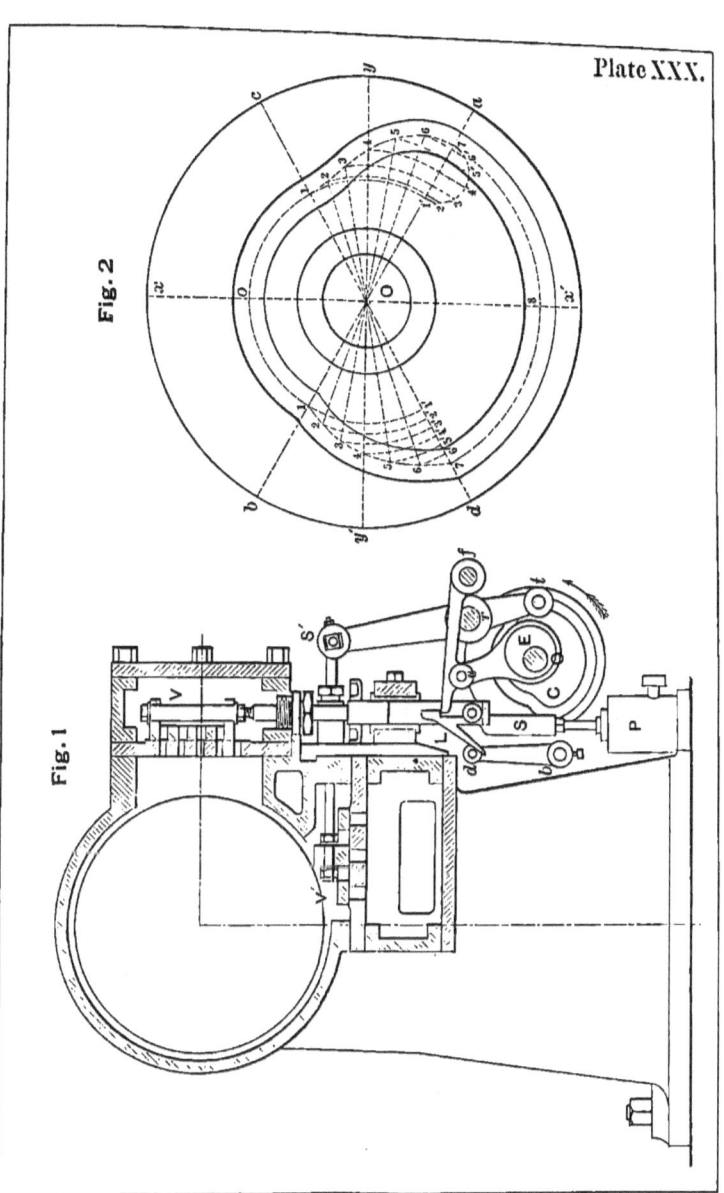

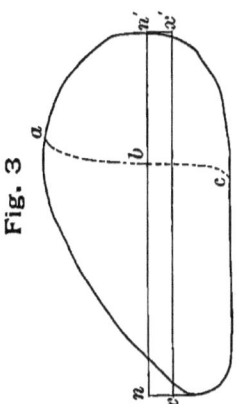

Plate XXXI.

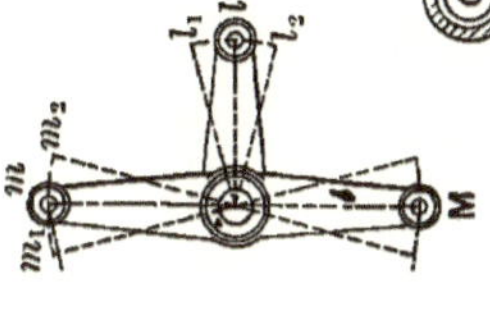

Fig. 2

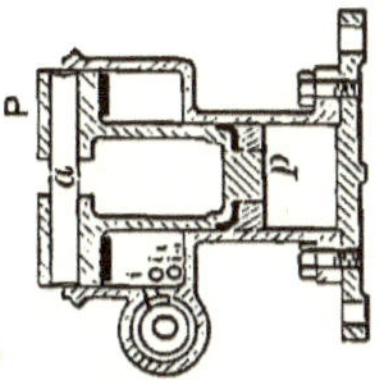

Fig. 3

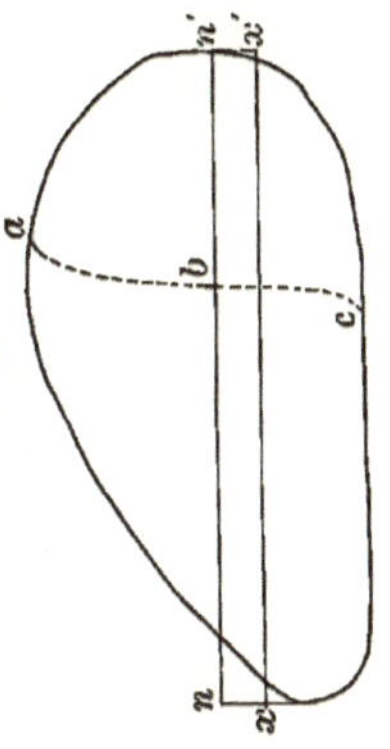

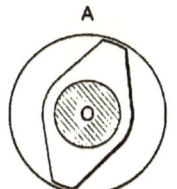

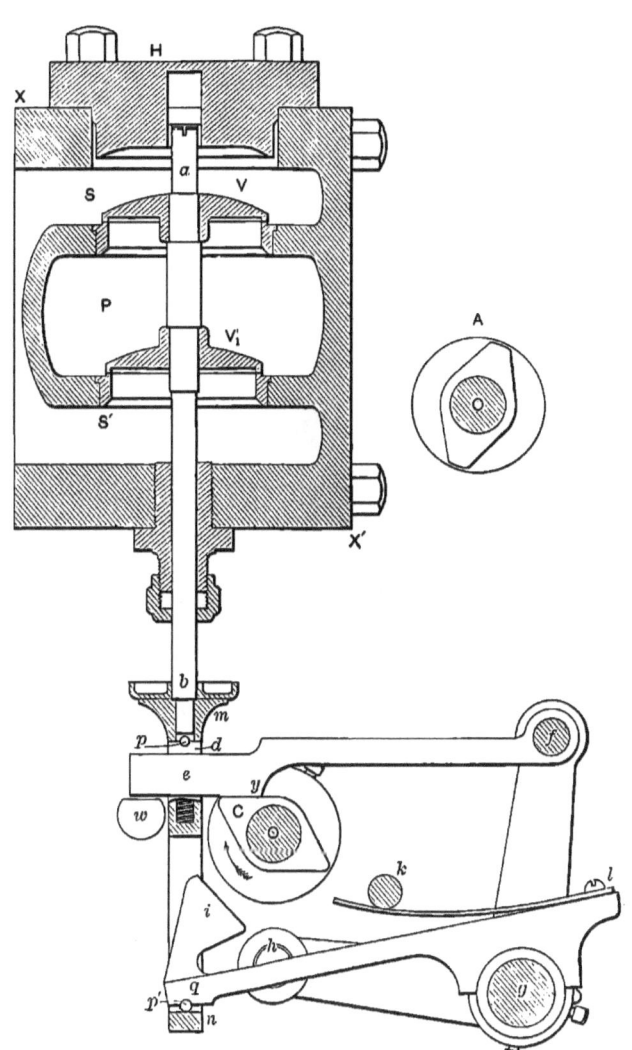

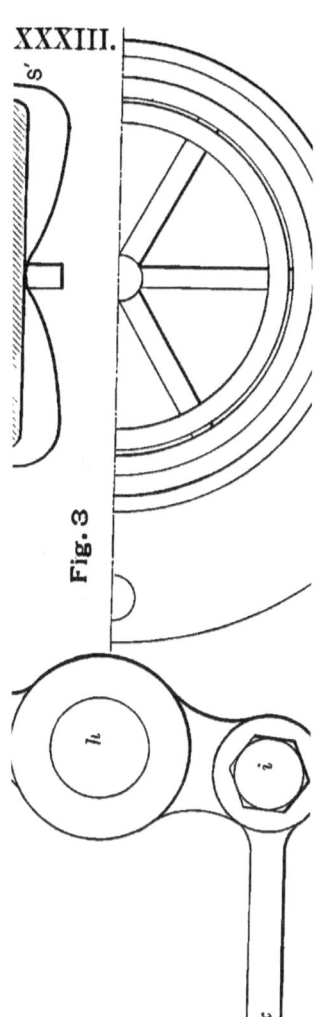

Fig. 3

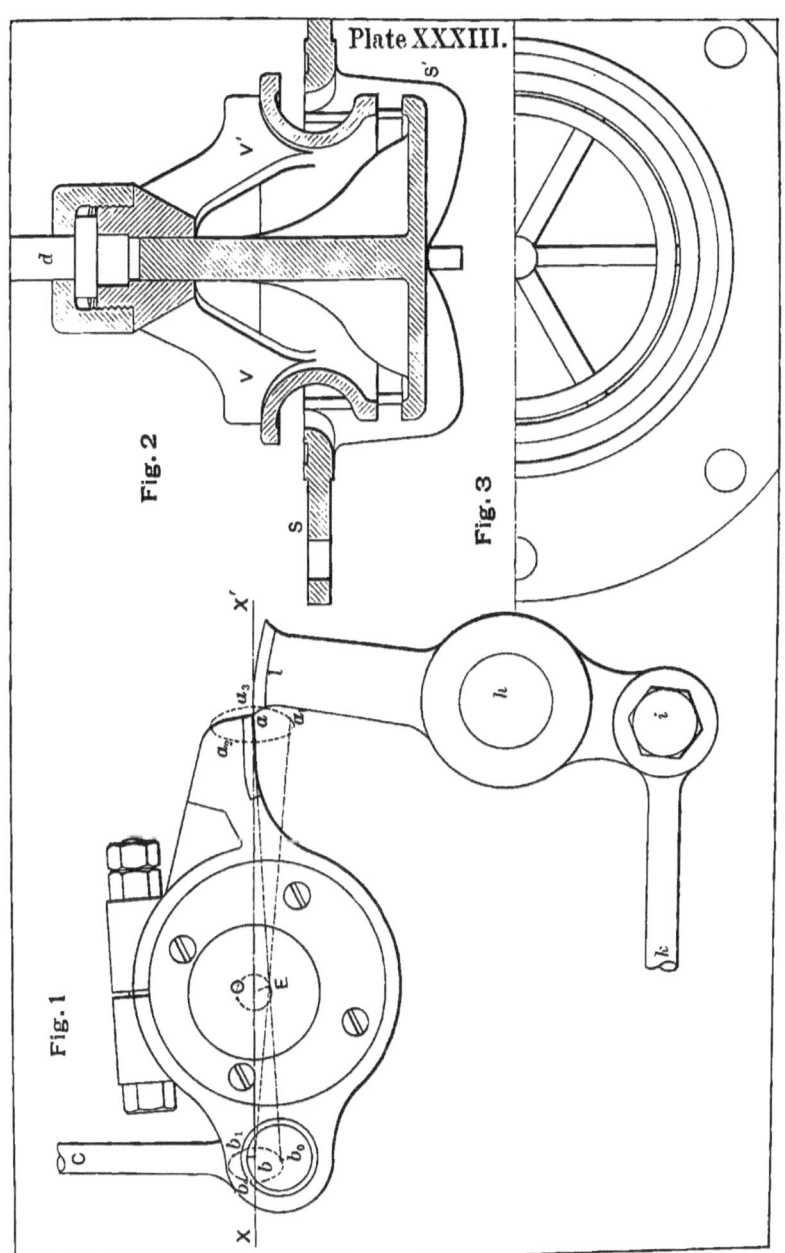

www.ingramcontent.com/pod-product-compliance
Lightning Source LLC
Chambersburg PA
CBHW032139230426
43672CB00011B/2390